T0174392

CRC Mathematical Modelling Series

Titles included in the series:

The Sentinel Method and Its Application to Environmental Pollution Problems

Jean-Pierre Kernévez

CRC Press

Boca Raton   New York

Acquiring Editor: *Tim Pletscher*
Senior Project Editor: *Susan Fox*
Cover Design: *Denise Craig*
PrePress: *Kevin Luong, Carlos Esser, Greg Cuciak*
Marketing Manager: *Susie Carlisle*
Direct Marketing Manager: *Becky McEldowney*

Library of Congress Cataloging-in-Publication Data

Kernevez, Jean Pierre.
　The sentinel method and its application to environmental pollution
problems / by Jean-Pierre Kernevez.
　　p.　cm. — (CRC mathematical modelling series)
　Includes bibliographical references and index.
　ISBN 0-8493-9630-1 (alk. paper)
　1. Water—Pollution—Mathematical models.　2. Least squares.
　3. Inverse problems (Differential equations) I. Title.
　II. Series.
　TD423.K43　1996
　628.1′68′015118—dc20

96-34576
CIP

© 1997 by CRC Press, Inc.

No claim to original U.S. Government works
International Standard Book Number 0-8493-9630-1
Library of Congress Card Number 96-34576
Printed in the United States of America　1　2　3　4　5　6　7　8　9　0
Printed on acid-free paper

Notations
Introduction

Notations

A $= -\Delta + \sigma I$: diffusion -reaction operator acting on the concentrations that are zero on Γ (or which have a null flux through Γ)

B   linear or nonlinear operator from V to H, transforming v in z, and defining the direct problem $v \rightarrow z$

$B'(v)$ derivative of $B(v)$

Cy   observation of the state y. $Cy(v) = B(v)$. Cy is the restriction of y to the observatory ω

     $= L^2(] \, 0,T \, [\times R^M)$ space of observations z and of sentinels w, m-tuples of square-integrable functions on $]0,T[$, endowed with the inner-product

$$(\varphi, \psi)_H = \sum_{k=1}^{M} \int_0^T \varphi(x_k, t) \psi(x_k, t) dt$$

Id_V identity application in V

J   cost function

$L^2(0, T)$, square-integrable functions on $] \, 0, T[$

$L^2(\Omega)$ square-integrable functions on Ω

M   number of observation points

N_1 number of sources of pollution. Number of pollution parameters

N_2 number of zones of a partition of Ω. Number of missing parameters

N $= N_1 + N_2$: total number of parameters, either pollution parameters like λ_i or missing parameters such as τ_j.

R^N space in which are the parameters $v = (\lambda, \tau)$

$$\begin{cases} v_i = \lambda_i & \text{for } 1 \le i \le N_1 \\ v_{N_1+j} = \tau_j & \text{for } 1 \le j \le N_2 \end{cases}.$$

R^M space in which is the observation z(t) at each time t.

S(z)   value of the sentinel S for observation z:

T   length of time - interval

V	$= R^N$.
W	pseudo - inverse of B, solving the inverse problem $z \to v$
a_i	i-th source point $(1 \le i \le N_1)$.
$\vec{c}$	pore water velocity, $[LT^{-1}]$
grad y	gradient of y
s_i	i -.th hat function, represented by a straight line on each elementary interval

$(t_{j-1}, t_j), 2 \le j \le N, 0 = t_1 < t_2 < ... < t_N = T$, and $s_i(t_j) = \begin{cases} 1 \text{ if } i = j \\ 0 \text{ if } i \ne j \end{cases}$

e^i	i-th vector of the canonical base of R^N: $e^i_j = \delta^i_j = \begin{cases} 1 \text{ if } i = j \\ 0 \text{ if } i \ne j \end{cases}$
t	time
w^i	i- th sentinel. $w^i \in H$

$x = (x_1, x_2)$ space coordinates in dimension 2

x_k	k-th point of observation. $(1 \le k \le M)$
y	pollutant concentration
y(x, t)	pollutant concentration at point x and time t
y(v)	state when the vector of parameters is v
y'	$= \dfrac{\partial y}{\partial t}$ partial derivative of y with respect to time
z	observation $z \in H$. $z = \left(z_1(t), \cdots, z_M(t) \right)$
$z_k(t)$	observation at point x_k at time t
y_0	concentration of a pollutant species at initial time t=0
div $\vec{q}$	divergence of $\vec{q}$
Δy	$= \text{div(grad y)}$ Laplacian of y

$\Lambda(v) = B'^*(v)B'(v)$ operator from $R^{N_1} \times R^{N_2} = R^N$ into itself ($N = N_1+N_2$).

∇y gradient of y

Λ_j^i $= \left(C\psi_i, C\psi_j\right)_H$, ($1 \le i, j \le N$), general term of matrix Λ

Ω open set of R^n, n= 1, 2 or 3, domain of the aquatic system

Ω_i.   i-th element of a partition of Ω.in zones

α vector of N_1 Lagrange multipliers α_i, $1 \le i \le N_1$

β vector of N_2 Lagrange multipliers β_j, $1 \le j \le N_2$

γ $\gamma = (\alpha, \beta)$

δ_j^i, $\delta_{i,j} = \begin{cases} 1 \text{ if } i = j \\ 0 \text{ if } i \ne j \end{cases}$ Kronecker symbol

$\delta(x-a)$ Dirac mass at point a

$\dfrac{\partial y}{\partial n}$ $= \vec{n}.\nabla y$, $\vec{n}$ unit outward normal

λ vector of the N_1 pollution intensities λ_i, $1 \le i \le N_1$

ρ state

τ vector $(\tau_1, ..., \tau_i, ..., \tau_{N_2})^T$, of the N_2 parameters τ_j defining the initial pollution on Ω,

ψ_i State of the system when all the parameters are equal to 0, except the i-th one, equal to 1

χ_j characteristic function of j-th zone, Ω_j. $\chi_j(x) = \delta_j^i$ if $x \in \Omega_i$

ω observatory, set of M points of measurement x_k in Ω.

Introduction

1 Physical motivation

This book presents the sentinels method, an implementation of the least squares method particularly well adapted to the identification of parameters in distributed ecosystems.

We give applications of the sentinels method to identification problems known as inverse problems in the mathematical-physics literature. An inverse problem can be defined as a problem where one has to estimate the cause from its effect, whereas a direct problem is concerned with the way to obtain the effect from the cause. We are interested with such inverse problems as those posed when there are incomplete data (initial and/or boundary conditions and/or some sources of pollution for example) in distributed ecosystems.

The modelling involves partial differential equations that govern the concentrations of pollutant species. Many such models can be found in the literature. See for example M.P. .Anderson and W.W. Woessner [1], W. Kinzelback[1], P. .Melli and P. Zannetti[1], A. Okubo [1], C. Taylor, and J.M.Davis, [1], C.L. Wrobel and C.A. Brebbia [1]).

There are important inverse problems in the area of geophysical applications, where they were studied initially. They would appear to be equally, if not more, important in the newer application areas of environmental systems, particularly in water resources. Diffusion equations arise, for instance, in stream pollution problems as well as in underground water flow problems. The same equations arise in air quality problems. These models would appear to be well enough founded to attempt system identification. In fact, with ever-increasing feasibility in high speed and low cost of digital computation, the scope of such inverse problems in modeling and simulation is bound to widen even further.

Many environmental problems contain incomplete data, among others in the initial conditions or in boundary conditions.

This situation is classical in Meteorology, where the initial data are never completely known. For other situations, already in 1795 Legendre and Gauss had seen the question. The environment systematically leads to questions of this kind. For examples, in pollution matters it may be difficult to know from where and in what quantity the pollution comes, or it can be voluntarily dissimulated by the polluter.

The question then is how to find the incomplete data? This is the kind of inverse problem we are faced with. We consider a distributed (eco system (surface water (river, estuary, bay, gulf or ocean), underground water, or atmosphere). We have at our disposal the pollutant concentration measured at a few points. The problem then is, by using these data is it possible to identify the parameters that control the behavior of the system. Examples of such parameters are the quantity of pollutant dumped by each source, or the overall quantity, dispersion coefficients, the position (and the discharge rate) of each source of pollution. It appears from the examples treated in this book that the answer to this question is, yes! : provided the data are rich enough, they enable us to identify some of the unknown parameters.

So the physical motivation of this book is the aim to implement the least squares method for identification of pollution in distributed ecosystems.

Note that the sentinel method is not the only one! Among others, for the determination of missing or incomplete initial conditions, we have the so-called assimilation methods which are least squares methods using slow or inertial manifolds. (Le Dimet [1]).

2 Historical background

The method for parameter identification is the familiar least squares technique. As said above, already in 1795 Legendre and Gauss had dealt with inverse problems and introduced the least squares method. In 1805 Legendre published his *Nouvelles méthodes pour la détermination des orbites des comètes*, in which he presented the least squares method.

This idea has since often been taken, to identify parameters in the case of systems governed by ordinary differential equations, then for distributed systems. The study of inverse problems in differential equations is less than 50 years old with a real explosion only in the last 10 or 20 years.(H.W. Engl and W. Rundell [1]). The application of least squares to distributed systems is based upon the theory of control of distributed systems (J.L. Lions [1], G. Chavent [3,4]).

Is it possible to do better?, faster?

This is the question J.L. Lions asked in about 1988-1992, by trying to see whether it was possible to calculate, in a problem where some data are missing, those which are useful, without calculating those it is useless to know. For example, in a river basin, where both the localizations and the quantities of pollution and the initial data are missing, is it possible to calculate the former without dealing with the latter?.

This led to the theory of sentinels, presented, with a certain number of variants, in Lions' book on this subject (J.L.Lions [10]). The method of sentinels has been placed in position for the theory by J.L. Lions in two articles in Comptes Rendus de l'Académie des Sciences in 1988 (J.L. Lions [2]) and has formed the subject, since then, of various developments (J.L. Lions [3-9]). But theory is not sufficient. It is necessary to have calculation algorithms applied to real cases. It is this second step, absolutely essential, of the sentinels program that is the motivation of this book.

3 Description of the method

We consider problems where a system depends upon parameters and is observed, with the aim of finding the parameters knowing this observation. It is a so-callled inverse problem, the direct problem being the one dealing with the determination of the observed quantities knowing the parameters.This book is restricted to environmental problems where the pollutant concentration is observed at some points x_k, $1 \leq k \leq M$. The k-th point of observation provides a measurement $z_k(t), 0 \leq t \leq T$. The direct problem is defined by

$$z = Bv \tag{1}$$

where v is a given vector of parameters in R^N and z is a calculated observation in $H = (L^2(0,T))^M$.

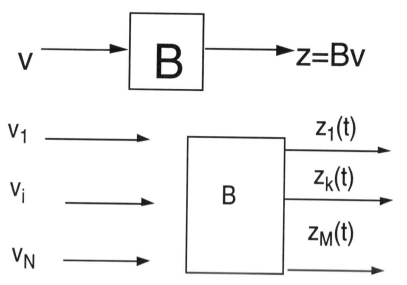

The space H of observations is equipped with the inner product
$$(\varphi, \psi)_H = \sum_{k=1}^{k=M} \int_0^T \varphi_k(t)\psi_k(t)dt \text{ and the associated norm}$$

$$|\varphi|_H = \sqrt{(\varphi, \varphi)}_H.$$

The inverse problem consists, given an observation z_d (the data), of minimizing $J(v)$ where
$$J(v) = \frac{1}{2}|Bv - z_d|_H^2 \tag{2}$$

It is well known that if u minimizes J then we have

$$\Lambda u = B^* z_d \tag{3}$$

where the matrix Λ is defined by

$$\Lambda = B^* B \tag{4}$$

The i-th row and j-th column entry of this matrix is defined by

$$\Lambda_i^j = \left(\Lambda e^i, e^j\right)_V = \left(Be^i, Be^j\right)_H$$

where e^i is the i-th vector of the canonical basis of $\mathbb{R}^N$.

Provided B is one to one then the matrix Λ^{-1} exists and (3) can be rewritten as

$$u = \Lambda^{-1}B^*z_d.$$

Then the i-th component of the vector u is

$$u_i = \left(u, e^i\right) = \left(e^i, \Lambda^{-1}B^*z_d\right) = \left(w^i, z_d\right)$$

where we define the sentinel w^i by

$$\Lambda v^i = e^i \tag{5}$$

$$w^i = Bv^i \tag{6}$$

and (3) can be rewritten as

$$u_i = \left(w^i, z_d\right)_H. \tag{7}$$

The sentinel method to identify u_i then consists of two steps:

1 First determine w^i, which is the so-called sentinel attached to the i-th parameter,
2 Then for each given observation z_d, calculate (7).

In conclusion, the sentinel to monitor the i-th parameter is defined by (5) and (6). Once it has been calculated, it can be employed by (7) to update the i-th component of the least squares estimation u of v.

Once w^i has been calculated, it can be used again and again for each new measurement and the calculation of u_i then is very fast. This may be of interest for early detection of situations of crisis in environmental systems.

A point of our approach is that it separates the more general and abstract structured aspects of the problems from the technicalities of the particular partial differential equation set-up involved.

Another point is that our appoach retains the similarity to the familiar finite-dimensional formulation as much as possible.

We see that all rests upon how the operator B acts on the parameters v in order to give the observation z by (1) and how the linear system (5) is solved.

Ultimately, everything rests upon the underlying distributed system.This book gives examples and develops applications, the model being described by the evolution equation

$$y' + Ay = \sum_{i=1}^{i=N_1} \lambda_i s_i(t)\delta(x - a_i) \tag{8}$$

where y is the concentration of pollutant, y' its time derivative, A is a second order elliptic operator, and in the second member

$$\begin{cases} \lambda_i = \text{unknown intensity} \\ s_i(t) = \text{known modulation} \\ \delta(x - a_i) = \text{Dirac mass at point } a_i \end{cases} \tag{9}$$

Let us rapidly travel through the ten chapters and the appendix. The chapters 1, 2, 3, 4, 5, 9, and 10 deal with linear cases (situations where the application : parameters→ observation is linear. The chapters 5, 6, 7, and 8 deal with nonlinear cases (situations where the application : parameters→ obsrervation is nonlinear. The contents of the different chapters are the following:

Chapter 1 deals with an aquifer. This study is motivated by a real case where pollution of underground water is due to two plants and the problem is to determine the responsability of each one.

Chapter 2 deals with pollution in a lake and some aspects of the employed numerical methods.

Chapter 3 considers the special case of dispertion of pollutant in a rectangular domain. This is motivated by the fact that in this case an exact solution can be obtained for the dispersion equation.

Chapter 4 does the same for a river, where a pollutant is subject to convection in plus of dispersion.

Chapter 5 is a transition between linear and nonlinear problems. It is important to note that when we say that a problem is linear, we mean that it is the application parameters→observation which is linear. In this Chapter 5 we consider a very simple system, whose evolution as a function of time is linear with respect to two parameters and nonlinear with respect to a third one. We first suppose known this third parameter, so that our parameter identification problem is linear. Then we suppose that the three parameters are unknown and we show in this example how successive linearizations enable to identify the three parameters.

Chapter 6 deals with a more complex system. It is a system subject to dispersion of pollutant and nonlinear reaction.

Chapter 7 also deals with a system distributed in space. Here the unknown parameters are dispersion coefficients, that we suppose constant by element.

Chapter 8 shows how the formalism of sentinels enables us to solve the problem to identify the position of a source of pollution. The potential applications are obvious.

Chapter 9 establishes convergence results which justify our approach of the problems by using finite dimensional models.

Chapter 10 introduces to the application of the method of sentinels to shallow waters.

An Appendix, written by J.L. Lions, returns to the situations studied in chapters 1 and 2 to bring two types of complements:

(1) a duality formula

(2) the introduction of functional spaces necessary for understanding the problem when one has no information about the initial data.

1 Identification of pollution in an aquifer

This chapter presents the sentinels for a linear case. Linear here means that the observation, when exactdepends linearly upon the parameters. As a first example we consider an underground aquifer.

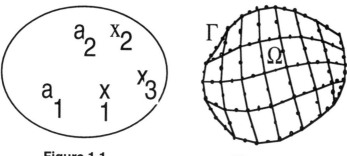

Figure 1.1   **Figure 1.2**

Two sources of pollution at points a_1 and a_2 three sensors at points x_1, x_2 and x_3 ; On the right a finite element mesh.

As a first example of the situations we may be faced to, let us consider (Figure 1.1) an aquifer that provides drinking water to the inhabitants of a region, but with a serious pollution in the wells. Two plants, respectively located at points a_1 and a_2, are potential sources of pollution, but each one places the responsibility on the other. Fortunately, we have data at our disposal. These data are the measurements of pollution during a period of time T, at three points denoted x_1, x_2 and x_3.

We are asked to identify the contribution of each source to the global pollution. This problem is motivating:

A first motivation is that groundwater pollution is a major issue. The assessment of the responsabilities of different sites a_1 and a_2 is legally interesting.

A second motivation is that it is very interesting to know the flow rate of each source of pollution. It enables us to simulate the plume and thus calculate accurate predictions (C. Poulard, R. Mose, and Ph. Ackerer [1]).

1.1 Modelling pollution transport in an aquifer

1.1.1 State equation

The pollutant behaves as a tracer and therefore, the problem is linear.

A two-dimensional model of an aquifer is possible because the horizontal dimensions of the aquifer are generally much larger than their depth, and the flow is sub-horizontal. The pollutant concentration y in an aquifer is governed by a transport equation expressing the balance of masses:

$$\frac{\partial y}{\partial t} = \frac{\partial y}{\partial t}_{\text{advection}} + \frac{\partial y}{\partial t}_{\text{dispersion}} + \{sources\}$$

The advection-dispersion equation for the concentration $y(x, t)$ is

$$\frac{\partial y}{\partial t} + \vec{c} \cdot grad(y) - div\big(D(x)grad(y)\big) = \sum_{i=1}^{N_1} \lambda_i s_i(t)\delta(x - a_i)$$

(Bear, 1972) (1.1.1)

Let Ω, of boundary Γ denote the water field. The pollutant is discharged in the aquifer by N_1 sources. The initial concentration of pollutant is supposed to be piecewise constant and denoted τ_j for the j-th of N_2 zones partitioning the domain W. We note:

- $N = N_1+N_2$,
- $x = (x_1, x_2)$ space coordinates, [L],
- t: time [T],
- y: pollutant concentration, $[ML^{-3}]$,
- $\vec{c}$: pore water velocity, $[LT^{-1}]$,
- D: dispersion tensor, $[L^2T^{-1}]$, D(x) represents the permeability of the water field, supposed to be isotrope,
- a_i: the i-th source of pollution ($1 \leq i \leq N_1$),
- λ_i: the intensity of its flow rate and
- $s_i(t)$ gives the shape of this flow rate as time varies.

We suppose that the flow problem has been solved as time varies in the interval]0, T[.

The evolution of the pollutant concentration y in Ω and on Γ is governed by the partial differential equation (1.1.1) and boundary and initial conditions.

1.1.2 Boundary conditions

$$\frac{\partial y}{\partial n} = 0 \text{ (Neumann boundary condition)} \qquad (1.1.2)$$

1.1.3 Initial conditions

The initial concentration in the water field is supposed to be piecewise constant on a partition $\{\Omega_j\}_{1 \leq j \leq N_2}$ of Ω.

$$y(x,0) = \sum_{j=1}^{j=N_2} \tau_j \chi_j(x) \text{ (initial condition), where}$$

$$\chi_j(x) = \begin{cases} 1 & \text{if } \in \Omega_j \\ 0 & \text{else} \end{cases} \tag{1.1.3}$$

For example, this partition of Ω could be a finite element mesh, as shown in figure 1.2.

The conditions (1.1.1), (1.1.2) and (1.1.3) can be expressed in a more concise form:

$$\begin{cases} y' + Ay = \sum_{i=1}^{N_1} \lambda_i s_i(t)\delta(x - a_i) \\ y(0) = \sum_{j=1}^{N_2} \tau_j \chi_j(t) \end{cases} \tag{1.1.4}$$

where

$$Ay = \vec{c} \cdot grad(y) - div\big(D(x)grad(y)\big)$$

1.1.4 Observation of pollutant concentration

The concentration is observed at M points x_k, $k = 1, ..., M$, which we call the observatory. We suppose that the available data are continuous time observation of the pollutant concentration at each of these M observation points. The observation consists of these M functions of time $z_d^k(t)$, $k = 1, ..., M$, $t \in]0,T[$. Let z_d denote these observed concentrations. The observation should be $y(x_k,t)$. It is so when the observation is exact. In that case let C denote the operator that, applied to a function such as the solution $y(x, t)$ of the state equations (1.1.1), (1.1.2) and (1.1.3), gives the M functions of time $y(x_k,t)$, $k = 1, ..., M$, $t \in]0,T[$. Actually, these measurements are perturbated by $S_1 = \Gamma_1 \times]0,T[$

$$z_d = Cy + \text{perturbation} \tag{1.1.5}$$

The perturbation is, in particular, due to measurement noise. We say that the observation is exact if in (1.1.5) the perturbation is zero. In that case, due to the linear dependence of the observation z upon the parameters λ_i and τ_j, we can write in matrix form the relation between an input v and an output z:

$$z = Bv \tag{1.1.6}$$

where v is the vector containing the input signal (the intensities of the pollutant sources and the initial conditions), z contains the calculated output signal, and B is an operator, which is determined as indicated herein by the equations that govern the system. Solving the direct problem consists of calculating z by using (1.1.6). Solving the inverse problem consists in estimating v by using the measurements z_d of z and applying the least squares method.

1.1.5 Space H of observations

The observation z calculated by (1.1.6), as well as the measured observation z_d, lie in a space H in which each element consists of M functions of time, one for each measurement point. This space H is equipped with the inner product

$$(\varphi, \psi)_H = \sum_{k=1}^{k=M} \int_0^T \varphi_k(t) \psi_k(t) dt \tag{1.1.7}$$

and the associated norm

$$|\varphi|_H = \sqrt[3]{(\varphi, \varphi)}_H \tag{1.1.8}$$

1.1.6 Unknown parameters

The origin of pollutant is unknown, whether it comes from a discharge in Ω or from an initial pollutant concentration in Ω. Therefore the unknown parameters of the system are N_1 pollution parameters

$$\lambda = (\lambda_1, ..., \lambda_j, ..., \lambda_{N_1})$$

and N_2 missing parameters:

$$\tau = (\tau_1, ..., \tau_j, ..., \tau_{N_2}).$$

We call pollution parameters those that appear in the right hand side of equation (1.1.1) and are of interest. We call missing parameters those that appear in the right-hand side of the initial condition (1.1.3) and are not interesting.

Despite that difference in nature of both types of parameters, we concatenate them, thus obtaining the vector v of all the parameters

$$v = (\underbrace{\lambda_1 ..., \lambda_i, ..., \lambda_{N_1}}_{N_1}, \underbrace{\tau_1 ..., \tau_j, ..., \tau_{N_2}}_{N_2}) \text{ , of length } N = N_1 + N_2. \tag{1.1.9}$$

In the following we call V the space R^N of parameters. Of course, only those elements of V with positive components have physical significance. The functions $s_i(t)$ and $\chi_j(x)$ are known.

We define the functions $\psi_i(1 \le i \le N_1)$ and $\varphi_i(1 \le j \le N_2)$ by

$$\begin{vmatrix} \psi_i' + A\psi_i = s_i(t)\delta(x - a_i) \\ \dfrac{\partial \psi_i}{\partial n} = 0 \qquad \text{for } 1 \le i \le N_1 \text{ and} \\ \psi_i(x, 0) = 0 \end{vmatrix} \tag{1.1.10}$$

$$\begin{cases} \varphi_j + A\varphi_j = 0 \\ \dfrac{\partial \varphi_j}{\partial n} = 0 \qquad \text{for } 1 \le j \le N_2 \\ \varphi_j(0) = \chi_j \end{cases} \tag{1.1.11}$$

For convenience we let

$$\Psi = (\underbrace{\psi_1 ..., \psi_i, ..., \psi_{N_1}}_{N_1}, \underbrace{\varphi_1 ..., \varphi_i, ..., \varphi_{N_2}}_{N_2}). \tag{1.1.12}$$

where $\psi_{N_1+j} = \varphi_j$ for j such that $1 \le j \le N_2$ and With the definitions (1.1.9) of v and (1.1.12) of Ψ the state y(v) solution of the system (1.1.4) can be expressed as

$$y(v) = \sum_{i=1}^{i=N} v_i \Psi_i = \Psi v \tag{1.1.13}$$

where we view Ψ as the row vector of functions ψ_i and v as the column vector of components v_i.

To each vector v of parameters the system (1.1.4) associates a state y, which, due to the linearity of this system, is a linear combination of the N elementary solutions ψ_i.

Applying (1.1.13) with $v = e^i$, i-th vector of the canonical basis of V, we have $\psi_i = y(e^i)$. Consequently the calculated observation for this vector of parameters v is

$$z(v) = Cy(v) = \sum_{i=1}^{i=N} v_i C\psi_i = C\Psi v = Bv \qquad (1.1.14)$$

where the operation parameters $v \rightarrow$ observation z is given by

$$B = C\Psi \qquad (1.1.15)$$

1.2 A sentinel attached to each parameter

Once a model is defined, it enables us to define the operator $B: v \rightarrow z$ where z is the calculated observation corresponding to the parameter v. The noise that exists in every measurement prevents the observation z_d to be of the form $z_d = Bv$ with $v \in V$, i.e; to belong to the range of B. The best we can do is to associate with z_d a vector u that minimizes the distance between z_d and the range of B Figure 1.3). We define for that the cost function

$$J(v) = |Bv - z_d|_H^2 . \qquad (1.2.1)$$

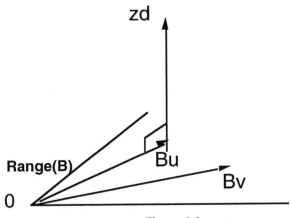

Figure 1.3

This cost function can be rewritten

$$J(v) = (\Lambda v, v)_V - 2(B^* z_d, v)_V + |z_d|_H^2 \qquad (1.2.2)$$

where Λ is the $N \times N$ matrix

$$\Lambda = B^*B \tag{1.2.3}$$

We essentially suppose B one-to-one. Then it is well-known that Λ is strictly positive definite, Λ^{-1} exists and there exists one and only one $u \in V$ minimizing J. It is also the solution of the linear problem

$$\Lambda u = B^* z_d \tag{1.2.4}$$

$$u = \Lambda^{-1} B^* z_d = W z_d \tag{1.2.5}$$

with

$$W = \Lambda^{-1} B^* \tag{1.2.6}$$

Thus u is a linear continuous function of z_d. The operator W can be called the pseudo-inverse of B. Indeed $WB = Id_V$ (the identity operator in V). We have $\forall v \in V$,

$$(u, \Lambda v)_V = (z_d, Bv)_H. \tag{1.2.7}$$

Taking $v = \gamma^i$ solution of $\Lambda v = e^i$, i-th vector of the canonical basis of V, (1.2.7) can be rewritten as

$$u_i = (z_d \cdot B\gamma_i)_H \tag{1.2.8}$$

so that if

$$w^i = B\gamma^i \tag{1.2.9}$$

then the i-th component of the minimizer u of J is given by the inner product

$$u_i = (z_d, w^i)_H. \tag{1.2.10}$$

where w^i is called the sentinel associated with the i-th parameter of the system.

Proposition 2.1

Let B: V → H be one to one from V to range(B). Then:

1. With the i-th parameter we can associate a sentinel w^i by $w^i = B\gamma^i$ where γ^i is the solution of the linear system, $\Lambda\gamma^i = e^i$.

2. The i-th component of the minimizer u of J is given by (1.2.10).

How does all that apply to our polluted aquifer? How do we pass from a problem relative to a distributed system where the state lives in an infinite-dimensional space to a finite-dimensional problem? The fact indeed is that the state is known if we know the N-dimensional vector v of parameters. The state is solution of (1.1.4)

$$\begin{cases} y' + Ay = \sum_{i=1}^{N_1} \lambda_i s_i(t)\delta(x - a_i) \\ y(0) = \sum_{j=1}^{N_2} \tau_j \chi_j(t) \end{cases}$$

(1.2.11)

This solution is given by (1.1.13)

$$y(v) = \sum_{i=1}^{i=N} v_i \Psi_i = \Psi v$$

and the calculated observation of state is (1.1.14)

$$z(v) = Cy(v) = \sum_{i=1}^{i=N} v_i C\Psi_i = C\Psi v = Bv$$

(1.2.12)

So for the aquifer problem the operator B: parameters $v \rightarrow$ calculated observation z is given by

$$B = C\Psi$$

(1.2.13)

It is now easy to describe two algorithms for the determination of the n-th source sentinel.

1 2.1 Direct method

1. Determination of the response of the system to a single source :

$: Be^i, \ 1\le \ \le N$ by calculating $\psi^i : Be^i = C\psi^i$ (1.2.14)

2. Determination of the matrix Lor i and j from 1 to N, calculate the components

$$\Lambda_i^j = \left(\Lambda e^i, e^j\right)_V = \left(Be^i, Be^j\right)_H = \left(C\psi^i, C\psi^j\right)_H \tag{1.2.15}$$

which can be rewritten as

$$\Lambda_i^j = \sum_{k=1}^{k=M} \int_0^T \psi^i(x_k, t)\psi^j(x_k, t)dt \tag{1.2.16}$$

3. Determination of the vector γ^n. The vector γ^n is obtained by solving

$$\Lambda\gamma^n = e^n \tag{1.2.17}$$

4. Determination of the sentinel w^n

$$w^n = B\gamma^n \tag{1.2.18}$$

5. Identification of the n-th parameter: the n-th parameter is given by the very simple formula

$$\lambda_n = u_n = \left(z_d, w^n\right)_H \tag{1.2.19}$$

1.2.2 Indirect method

A method not requiring the calculation of the matrix Λ would be most welcome. The reason is that, if the matrix Λ is large, it occupies a lot of memory, and its employ necessitates much computer time. Solving (1.2.17) is equivalent to minimize the cost function

$$J(\gamma) = \frac{1}{2}(\Lambda\gamma, \gamma)_V - (e^n, \gamma)_V \tag{1.2.20}$$

Its value J(v) can also be written as

$$J(\gamma) = \frac{1}{2}(B\gamma, B\gamma)_H - (e^n, \gamma)_V \tag{1.2.21}$$

Minimizing (1.2.21) (without any constraint on γ) is equivalent to minimize

$$J = \frac{1}{2}(C\rho, C\rho)_H - (e^n, \gamma)_V. \tag{1.2.22}$$

under the constraint that the state ρ is related to the control $\gamma = (\alpha, \beta)$ by the equations

$$\begin{cases} \rho' + A\rho = \sum_{i=1}^{N_1} \alpha_i s_i(t)\delta(x - a_i) \\ \rho(0) = \sum_{j=1}^{N_2} \beta_j \chi_j(x) \end{cases} \tag{1.2.23}$$

Let $\rho(\gamma)$ be this state for the control vector $\gamma = (\alpha, \beta)$. We pass from $\gamma = (\alpha, \beta)$ to $C\rho(\gamma)$ in the same way we passed from $v = (\lambda, \tau)$ to $Cy(v)$, indicated in (1.1.14).

$$z(v) = Cy(v) = \sum_{i=1}^{i=N} v_i C\psi_i = C\Psi v = Bv$$

Here we have the observation of the state corresponding to the optimal control

$$C\rho(\gamma) = \sum_{i=1}^{i=N} \gamma_i C\psi_i = C\Psi\gamma = B\gamma \tag{1.2.24}$$

So an alternate way to calculate Bg ($\in H$), γ been given, is to solve for ρ the system of equations (1.2.23), which provides $\rho(\gamma)$, then to apply the operator C, thus obtaining $C\rho(\gamma) = B\gamma$

$$C\rho(\gamma) = C\Psi\gamma \tag{1.2.25}$$

Once the optimal γ has been determined, we have

$$w^n = B\gamma^n \tag{1.2.26}$$

1. Determination of the vector γ^n. The vector γ^n is obtained by solving the optimal control problem (1.2.22) and (1.2.23).

2. Determination of the sentinel w^n

$$w^n = C\rho(\gamma^n) \tag{1.2.27}$$

The indirect algoritm consists in determining the sentinel $w^n = B\gamma^n$ by solving the optimal control problem. There is no need to calculate Λ. This means that this method is interesting when there are many unknowns in τ. But, on the other hand, we have to solve the optimal control problem for each pollutant source. Minimizing J with a gradient method such as the conjugate gradient requires solving the state and adjoint state equations three times per iteration.

1.3 Examples of similar problems

By our model we mean a system whose state y obeys an evolution equation of the form

$$y' + Ay = \sum_{i=1}^{N_1} \lambda_i s_i(t) \delta(x - a_i) \qquad (1.3.1)$$

together with initial conditions of the form

$$y(0) = \sum_{j=1}^{j=N_2} \tau_j \chi_j(x). \qquad (1.3.2)$$

We wish to identify one of the parameters The parameters in λ and τ are unknown, but data are available. These are the measured concentrations z_d of y at M points x_k. These M points x_k constitute an observatory ω.

We have the balance of masses

$$\frac{\partial y}{\partial t} = \frac{\partial y}{\partial t}_{advection} + \frac{\partial y}{\partial t}_{dispersion} + \{sources\}$$

It is important to distinguish between pollution parameters λ_i and missing parameters τ_j. The former are easier to calculate and are interesting; the latter are more difficult to identify and, anyway, do not interest us.

1.3.1 Pollutant transport in shallow water

We model pollutant transport in shallow water by

1. an hydrodynamic model, where we suppose that, either by measurements, or numerical simulation, we know the averaged horizontal velocity $\vec{c} = (c_1, c_2)$

2. the transport equation

$$\frac{\partial y}{\partial t} + \vec{c}.\nabla y - div\big(D\ grad(y)\big) + \sigma y = \frac{1}{h} \sum_{i=1}^{N_1} s_i(t)\delta\big(x - a^i\big) \qquad (1.3.3)$$

where

y is the biochemical oxygen demand (BOD),

i is the discharge number,

$s_i(t)$ is the flow rate,

$\delta\big(x - a^i\big)$ is the Dirac mass at the discharge point a_i,

σ is a kinetic parameter,
D is a dispersion coefficient and
h is water height.

Let A be the operator defined by

$$Ay = \vec{c}.\nabla y - div\big(D\ grad\ (y)\big) + \sigma y \qquad (1.3.4)$$

We again adopt a piecewise constant initial pollutant concentration and have an equation similar to that already encountered for an aquifer :

$$y' + Ay = \sum_{i=1}^{N} \lambda_i s_i(t)\delta\big(x - a_i\big) \text{ (state equation)} \qquad (1.3.5)$$

$$y(0) = \sum_{j=1}^{j=N_2} \tau_j \chi_j(x) \text{ (initial condition)} \qquad (1.3.6)$$

and implicitly assume that operator A acts on functions with zero flux boundary condition:

$$\frac{\partial y}{\partial n} = 0 \text{ (boundary condition)} \qquad (1.3.7)$$

1.3.2 BOD transport in a Gulf

An open set Ω (the Gulf) has a boundary Γ composed of two parts:

1. the land $\Gamma 1 = \partial\Omega_N$ (the continent or the boundary of islands)

2. the part $\Gamma 0 = \partial\Omega_D$ that is exposed to the open sea, boundary on which we have a latitude. Exactly as in a harbor simulation, we have a large latitude in the choice of the part exposed to the open sea.

We then have **pollutions** we wish to estimate, which are lumped at points $\{a_1,...,a_{N_1}\}$ and are expressed by $\sum_{i=1}^{i=N_1} \lambda_i s_i(t)\delta(x-a_i)$, $s_i(t)$ known. We could take the point a_i varying with time: $a_i = a_i(t)$, if for example a shuttle is employed to measure continuously the contaminant concentration.

We also have **Perturbations** we cannot identify, which are lumped at points $\{b_1,...,b_{N_2}\}$ and are expressed by $\sum_{j=1}^{j=N_2} \tau_j g_j(t)\delta(x-b_j)$, $g_j(t)$ known, without any information about the τ_j. One could also take the point b_i varying with time: $b_i = b_i(t)$. Therefore

$$y' + ASy = \sum_{i-1}^{i=N_1} \lambda_i s_i(t)\delta(x-a_i) + \sum_{j=1}^{j=N_1} \tau_j g_j(t)\delta(x-b_j) \quad \text{on } Q = \Omega \times]0, T[$$

where $y' = \dfrac{\partial y}{\partial t}$ and $Ay = \vec{c} \cdot \text{grad}(y) - \Delta y + sy$.

We introduce:

1. An **observatory** ω in the Gulf, consisting of M points of measurement x_k, observation points often located on the coast, or in a neighborhood of the coast, $\omega = \{x_k\}_{1 \leq k \leq M}$.

2. A **sentinel** W, that is M functions of time $w_k(t)$ whose inner product with the M functions $y(x_k,t)$ provides the pollution parameter λ_i:

$$\sum_{k=1}^{k=M} \int_0^T w_k(t)y(x_k,t)dt = \lambda_i$$

Here the pollution parameters, those we are interested in, are the λ_i's, and those we are not interested in, are the missing data τ_j.

1.3.3 Pollutant transport in atmosphere

$$\frac{\partial y_i}{\partial t} + div(\vec{c}y_i) - div(D\ grad\ (y_i)) = f_i + S_i$$

Here, y_i is the concentration of the i-th pollutant among p species, i.e., $i = \{1, \cdots, p\}$i, $\vec{c}$ is the wind velocity field, D is the diffusivity tensor, $f_i(c_1, \ldots, c_p)$ is the chemical reaction term and S_i is the emission rate of the i-th species.

1.3.4 Monitoring the pollution in a river

$$\begin{cases} \frac{\partial y}{\partial t} + \vec{c}.\nabla y - D_1\frac{\partial^2 y}{\partial x_1^2} - D_2\frac{\partial^2 y}{\partial x_2^2} + \sigma y = f_1(t)\delta\left(x - a^1\right) + f_2(t)\ \delta\left(x - a^2\right) & in\ \ \Omega \times]0, T[\\ y(x, 0) = y_0(x) & in\ \ \ \ \Omega(at\ t = 0) \\ \frac{\partial y}{\partial v_D} = h & on\ \ \ \ \partial\Omega_N \times]0, T[\\ y = g & on\ \ \ \ \partial\Omega_D \times]0, T[\end{cases}$$

$$(1.3.8)$$

where

$\vec{c}$ is the velocity field of the river

D_1 and D_2 are constant dispersion coefficients

σ is a reaction coefficient

a^1 and a^2 are the location of two pollution sources

f_1 and f_2 are the intensity of these sources

y_0 is the initial concentration of pollutant

g is the concentration of pollutant on the upstream boundary $\partial\Omega_D$ of the river

$\frac{\partial y}{\partial v_D} = D_1\frac{\partial y}{\partial x_1}v_1 + D_2\frac{\partial y}{\partial x_2}v_2$, where $\vec{v} = (v_1,\ v_2)$ is the outer unit normal to Ω

h is the flux of pollutant caused by dispersion on the lateral and downstream boundary $\partial\Omega_N$

We make the hypothesis that all coefficients and the right-hand sides in (1.3.4.) are known, except f_1, f_2, y_0, g and h. Here, to change a little, we suppose that the observatory is an open set ω of W, $\omega \subset \Omega$, on which a measurement $z \in L^2(\omega \times]0, T[)$ of y is available. Here the space of observations is contained in $H = L^2(\omega \times]0, T[)$.

If y = 0 on $\omega \times]0, T[$, from the theorem of Misohata [1] we know that y is identically zero on $(\Omega - \{a^1, a^2\}) \times]0, T[$, hence y = 0 a.e. on $\Omega \times]0, T[$, which implies that f_1, f_2, y_0, g, and h are zero, so that B is injective.

Problem 3.1
We want to monitor the total amount of pollutant injected by sources one and two during the observation period [0,T[: $\int_0^T f_1(t)dt + \int_0^T f_2(t)dt$.

1.4 Flow rate

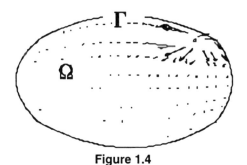

Figure 1.4

A point source of pollution and the pollutant dispersion flux at a given time.

Up to now we were supposing that every function s_i: $t \rightarrow s_i(t)$ was known. If it is unknown, is it possible to identify this function? We are going to examine this problem in the case of only one source, located at a known point a, and discharging a pollutant at a flow rate s(t) (kilograms per day). The present problem appears as a particular case of the one already treated, where all the sources a_i are located at the same point a and where the functions s_i constitute a truncated base of $L^2(]0,T[)$ (hat functions, for example, or piecewise constant functions). We identify the instantaneous flow rate of pollutant discharged in water by treating with the sentinels method the measured pollutant concentration at observation points. The numerical results prove the interest of that method, since the identification can be satisfactorily effected, provided, of course, the sensors are reasonnably placed with respect to the sources of pollution. The state y of the system is the solution of the Cauchy problem

$$\begin{cases} y' + Ay = f \\ y(0) = y_0 \end{cases} \tag{1.4.1}$$

where the dispersion-reaction operator A is defined by

$$Ay = -\Delta y + \sigma y \tag{1.4.2}$$

acting on functions y such that the source term f is of the form

$$f(x,t) = s(t)\delta(x-a) \tag{1.4.4}$$

where 2(x-a) is the Dirac mass at point a.

We suppose that the source function s is a linear combination of hat functions $t \to s_i(t)$ such as those represented Figure 1.5.

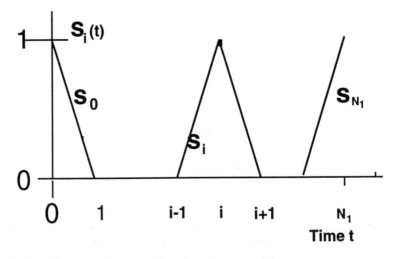

Figure 1.5

Function: $t \to s_i(t)$. Time t in abscissa from 0 to N_2 = 24h.

We partition the interval $]0,T[$ into N_1 elementary intervals $](i-1)\times\Delta t, i\times\Delta t[$, $1\le i\le N_1$ and define s on $(0,T)$ as

$$s(t)=\sum_{i=1}^{i=N}\lambda_i s_i(t) \tag{1.4.5}$$

where the so-called hat function s_i is affine on each elementary interval, and

$$s_i(j\Delta t)=\begin{cases}1 \ if \ j=i \\ 0 \ if \ j\ne i\end{cases} \tag{1.4.6}$$

This corresponds well to the situation often encountered in pratice, where one represents the flow rate as a function of time by a piecewise affine function

$$f(x,t)=\sum_{i=1}^{i=N_1}\lambda_i \ s_i(t)\delta(x-a) \tag{1.4.7}$$

$\sigma\times y$ is a first order reaction term (pollutant consumption).

The initial condition y_0 is the initial distribution of concentration. The initial concentration in the water field is supposed to be piecewise constant on a partition $\{\Omega_j\}_{1\le j\le N_2}$. For example, this partition of W could be a finite element mesh (Figure 1.6), provided the number of elements is not too large.

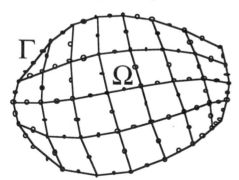

Figure 1.6

We suppose the initial concentration y_0 to be of the form

$$y_0(x) = \sum_{j=1}^{j=N_2} \tau_j \chi_j(x) \tag{1.4.8}$$

where

$$\chi_j(x) = \begin{cases} 1 \text{ if } x \in \Omega_j \\ 0 \text{ otherwise} \end{cases}$$

These functions χ_j constitute a truncated basis of $L^2(\Omega)$.

The system of equations (1.4.1) then can be rewritten as

$$\begin{cases} y' + Ay = \sum_{i=1}^{i=N_1} \lambda_i \; s_i(t)\delta(x-a) \\ y(0) = \sum_{j=1}^{j=N_2} \tau_j \chi_j(x) \end{cases} \tag{1.4.9}$$

We do not know the functions s: $t \rightarrow s(t)$ and y_0: $x \rightarrow y_0(x)$, that is neither the parameters λ_i nor the parameters τ_j. But we have available the measurement of y(x, t) at a finite number of points x_k, constituting the *observatory* $\omega = \{x_1, x_2, ..., x_M\}$. The problem is to calculate λ_i values knowing some noisy measurements $z_k(t)$ of $y(x_k, t)$. We are in the same situation as above and the only difference being that all the source points a_i now, are located at the same place a and the functions s_i are hat functions.

Therefore, we can employ the same algorithm.

1.5 Numerical experiments

Figure 1.7

Configuration involving four points of measurement and a point-wise source S, of intensity 0.32 (the intensities are here randomly generated between 0 and 10), and of type "city" $s(t) = 1 + \sin\left(\frac{2\pi}{T}t\right)$ $T = 24.$

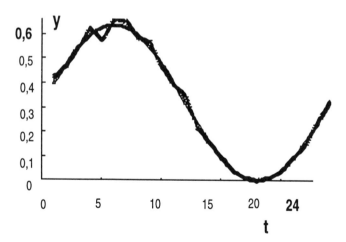

Figure 1.8

Noise of 20%. Identified source of pollution. The signal s(t) was a perfect sinusoid. The identified signal is a linear combination of hat functions of support 2h. It is very close to s(t).

Of course, we can use, instead of hat functions, other functions. For example we can look for an approximation of s by piece-wise constant functions:

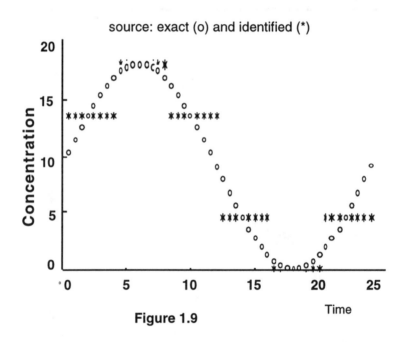

Figure 1.9

1.6 Sentinels and pseudo-inverse

The least squares method is the classical method of identifying parameters. It is the method we used in Section 1.2 to define the optimal vector of parameters u. It consists of minimizing a cost function.

$$J(v) = \frac{1}{2}\left|Cy(v) - z_d\right|_H^2 = \frac{1}{2}\left|Bv - z_d\right|_H^2, \ v = (\lambda, \ \tau) \tag{1.6.1}$$

Of course, variants are possible, in considering, for example, weights

$$J(v) = \frac{1}{2} \sum_{k=1}^{k=M} \int_0^T m_k(t) \big(y(x_k, t) - z_d(t) \big)^2 dt \tag{1.6.2}$$

N.B. Here both λ and τ are calculated. The classical least squares method consists of simultaneously calculating all the components of u, either in solving the linear system.

$$\Lambda u = B^* z_d \tag{1.6.3}$$

where Λ is a positive definite, symmetric matrix or in looking for v minimizing the cost function

$$J(v) = \frac{1}{2}(\Lambda v, v)_V - (B^* z_d, v)_V \tag{1.6.4}$$

It is not opportune to "compare" sentinels and pseudo-inverses since the sentinel w^n precisely is the n-th component of the pseudo-inverse W.

A sentinel (theoretically) enables one to calculate only the parameter one wishes and not the others. Traditionally, in the present case we calculate the N_1 components of λ without calculating τ (i.e., the N_2 last components of vector v).

An interesting property of the sentinels method compared to the classical least squares method is that possibility of using the sentinel w'' for a repetitive fast calculation of the parameter u_n. Once w'', the n-th component of W, has been calculated, it can be indefinitely used again and again, rapidly providing the value $u_n = (w^n, z_d)_H$, for each new observation z_d. We have seen that to obtain the n-th parameter by the sentinels method only needs the solution of an optimal control problem. Similarly the classical least squares method consists of solving the following optimal control problem: The state y is defined by

$$\begin{cases} y' + Ay = \sum_{i=1}^{i=N_1} \lambda_i \ s_i(t) \delta(x-a) \\ y(0) = \sum_{j=1}^{j=N_2} \tau_j \chi_j(x) \end{cases} \tag{1.6.5}$$

The control is $v = (\lambda, \tau)$. The cost to minimize is

$$J(v) = \frac{1}{2} \big| Cy(v) - z_d \big|_H^2 \tag{1.6.6}$$

A priori both the classical least squares method and the sentinels method need the matrix Λ, whose calculation may be a costly operation. This matrix has N^2 terms and requires solving the state equation N times. Let us remember that the discretization of an aquifer often requires far more than 10 000 elements! and so N_2, as we often defined it (the number of finite elements), is more than 10^4. Of course, it still remains possible to define super-elements, that is larger zones on which the initial concentration is supposed to be constant. But, above all, there is the indirect method.

2 Identification of pollution in a lake

2.1 Pollution of a lake

We consider in this chapter a water field polluted by a chemical species. The phenomena we have to take into account are dispersion and consumption of the pollutant. One may think of a lake polluted by biological oxygen demand (BOD). The physical problem is to identify the amount of pollutant discharged by each source. Measurements are available to achieve this goal. These are the pollutant concentrations measured at a few points[1], which we call the observatory.

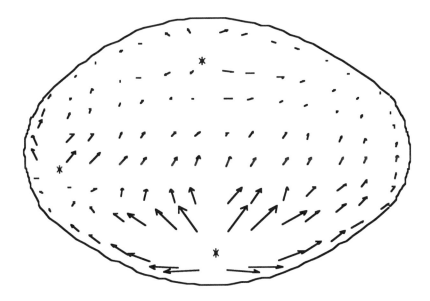

Figure 2.1

*A * indicates the position of a source of pollution. Arrows indicate the calculated flux of dispersion at a given time.*

---

[1]Because the sensors are expensive!

2.1.1 State equation

Let Ω be an open bounded set in R^2, of boundary Γ, smooth enough.[2,3] The pollutant is the BOD. The BOD concentration y is defined as the solution of the following partial differential equation, and boundary and initial conditions.

$$\begin{cases} \dfrac{\partial y}{\partial t} + Ay = \sum_{i=1}^{N_1} \lambda_i s_i(t)\delta(x - a_i) & \text{in } Q = \Omega \times]0, T[\\[2mm] \dfrac{\partial y}{\partial n} = 0 & \text{in } \Sigma = \Gamma \times]0, T[\\[2mm] y(x,0) = \sum_{j=1}^{N_2} \tau_j \chi_j(x) & \text{in } \Omega \end{cases} \qquad (2.1.1)$$

where

$$Ay = -\Delta y + \sigma \, y \, , \quad A = -\Delta + \sigma \, ,$$

i is the discharge number,

$\lambda_i s_i(t)$ is the flow rate of the i-th source

$\displaystyle\sum_{i=1}^{i=N_1} \lambda_i s_i(t)$ is the total flow rate,

$\delta(x - a^i)$ is the Dirac mass at this discharge point a_i,

the positive parameter σ characterizes a first order chemical reaction of disappearance. That is to say that the consumption of pollutant is of the form σy ,the points $a_i = (a_{i_1}, a_{i_2})$ $1 \le i \le N_1$ are located in Ω and are the sources of pollution

λ_i is the i-th source intensity of pollutant discharge.

$s_i :] \, 0, T[\rightarrow R$ is the shape of the discharge of the i-th source of pollution on a period of T hours (often T = 24 Hours).

---

[2]For example piecewise C^1.

[3]Although we restrict ourselves to examples in R^2 our methods apply equally well to examples in R^1 or R^3.

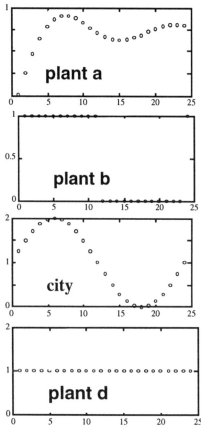

Figure   2.2

This figure shows four examples among the infinity of possible shapes of the daily source discharge.

Examples of functions $s_j.$, s_i values versus time. Time t in abscissa from 0 to 24 h. Daily pollutions.a) plant working all the day long, b) plant working half time, c) city polluting with a period of 24h, d) plant polluting at a constant rate.

2.1.2 Initial conditions

We define a partition of the spatial domain Ω in N_2 zones Ω_j (of a finite elements mesh, for example) in each of which we suppose the initial concentration to be constant. We suppose that the initial concentration is a linear combination of a finite number N_2 of functions χ_j:

$$y(x,0) = \sum_{j=1}^{N_2} \tau_j \chi_j(x)$$ (2.1.2)

where χ_j is the characteristic function of the j-th zone

$$\chi_j(x) = \begin{cases} 1 \text{ if } x \in \Omega_j \\ 0 \text{ otherwise} \end{cases}$$ (2.1.3)

2.1.3 Boundary conditions

The natural boundary condition is a null flux condition along Γ (the boundary is impervious to the pollutant):

$$\frac{\partial y}{\partial n} = 0 \text{ on } \Sigma = \Gamma \times] \, 0, T [$$ (2.1.4)

This is only to fix ideas. Alternatively, we could impose Dirichlet boundary conditions, for example

$$y = 0 \ on \ \Sigma = \Gamma \times]0, T[.$$ (2.1.5)

Let us denote

$$y' = \frac{\partial y}{\partial t},$$

τ_j is the initial condition on element Ω_j, $1 \leq j \leq N_2$.

$$\lambda = (\lambda_1, ..., \lambda_j, ..., \lambda_{N_1}) \qquad \tau = (\tau_1, ..., \tau_j, ..., \tau_{N_2})$$

$$N = N_1 + N_2,$$

$$v = (\underbrace{\lambda_1 ..., \lambda_i, ..., \lambda_{N_1}}_{N_1}, \underbrace{\tau_1 ..., \tau_j, ..., \tau_{N_2}}_{N_2}) = (\lambda, \tau)$$

Suppose we do not know the parameters v. In counterpart we have at our disposal $y(x_k, t; v)$ at M points x_k, the time history, as time t varies in the time interval $]0, T[$, of the pollutant concentration.

The problem we address to is to determine only one of the parameters in λ by using the information hidden in these observations. We are not interested in $^\tau$ (i.e.,. the initial condition).

The points x_k constitute what we will call an observatory $\omega = \{x_1, x_2, ..., x_M\}$. The observation z consists of M functions of time,

$$z_k : t \to z_k(t) = y(x_k, t; v) \qquad 1 \le k \le M. \tag{2.1.6}$$

In the following C is the exact state observation operator

$$Cy(x, t; v) = \{y(x_1, t; v), ..., y(x_M, t; v)\} \tag{2.1.7}$$

The intensities λ_i of pollution are called pollution parameters and the parameter values τ_j which define the initial pollution, are called missing parameters. We are interested in identifying the value of one or several pollution parameters λ_i. Once this parameter has been determined, we know the corresponding term of pollution $\lambda_i s_i(t)$, contribution of the i-th source to the global pollution. But we are not interested in identifying each parameter τ_j of the initial pollution. Rather we could be interested in identifying the initial total amount of pollution in Ω,

$$\int_\Omega y(x, 0) dx = \sum_{j=1}^{N_i} \tau_j \int_{\Omega_j} 1 dx. \tag{2.1.8}$$

A way to estimate this initial amount of pollutant in Ω will be shown in Section 4 of this chapter.

Remark 2.1.1

Should we desire to identify one or several coefficients τ_j we could follow the same method we are going to describe for a pollution parameter λ_i.

Remark 2.1.2

Of course, due to the physical origin of the problem the components of vector v and the functions s_j must be positive. The same applies for functions χ_j, from which it results that $y(v) \geq 0$.

2.2 Sentinels

2.2.1 Pseudo inverse

As in Chapter 1, it is a simple exercise to show that the cost function

$$J(v) = \frac{1}{2}\left|Cy(v) - z_d\right|^2_H, v \in V \tag{2.2.1}$$

is minimized by u, solution of

$$\Lambda u = B^* z_d \tag{2.2.2}$$

or equivalently

$$u = W z_d \tag{2.2.3}$$

where

$$W = \Lambda^{-1} B^*. \tag{2.2.4}$$

We have the following

Proposition 2.2.1

For every z_d in H there is one and only one vector u such that

$$\begin{cases} u \in V \\ J(u) \leq J(v) \; \forall v \in V \end{cases} \tag{2.2.5}$$

This vector u is the solution of the linear problem

$$\Lambda u = b \tag{2.2.6}$$

where Λ and b are given by

$$\Lambda = B^*B = \begin{bmatrix} (C\Psi_1, C\Psi_1)_H & \cdots & (C\Psi_1, C\Psi_N)_H \\ \vdots & \ddots & \vdots \\ (C\Psi_N, C\Psi_1)_H & \cdots & (C\Psi_N, C\Psi_N)_H \end{bmatrix} \tag{2.2.7}$$

and

$$b = B^*z_d = \begin{bmatrix} (C\Psi_1, z_d)_H \\ \vdots \\ (C\Psi_N, z_d)_H \end{bmatrix} \tag{2.2.8}$$

$$u = Wz_d \tag{2.2.9}$$

where

$$W = \Lambda^{-1}B^* \tag{2.2.10}$$

By (2.2.9) - (2.2.10) to each $z_d \in H$ there corresponds an element u *of* V. It is the solution of the least squares problem

$$\inf_V J(v) \tag{2.2.11}$$

where J(v) is defined by (2.2.1) and z_d is any element in H, in or not in the range of B. We easily check that

Proposition 2.2.2

$WB = Id_V$ (identity operator on V): $\tag{2.2.12}$

Proof

$\forall u \in V,$ *define* $z_d = Bu$ *and* $\tilde{u} = Wz_d . \tilde{u}$ minimizing $|Bv - Bu|_H , \tilde{u} = u.$

The operator W is called the pseudo-inverse of the operator B, or the generalized inverse of B. Since $WB = Id_V$, the operator W is also called the left inverse of B. We insist on the fact that the only requirement for W to exist is that B be injective and range (B) be closed in H. The latter holds since

V is finite dimensional of dimension N and injective from V into H, so that BV itself is an N-dimensional subset of H, and therefore is closed.

2.2.2 Sentinels

If e^i denotes the i-th vector of the canonical basis of V, the i-th component of u is given by

$$u_i = (e^i, u)_V = (e^i, Wz_d)_V = (W^* e^i, z_d)_H = (B\Lambda^{-1} e^i, z_d)_H = (w^i, z_d)_H \quad (2.2.13)$$

Definition   2.2.1

w^i *is the sentinel associated with the parameter* v_i.

Proposition   2.2.3

The i-th parameter is the inner product in H of the sentinel w^i and the observation z_d

$$u_i = (w^i, z_d)_H \tag{2.2.14}$$

where w^i is defined by

$$w^i = B\gamma^i \tag{2.2.15}$$

where γ^i is defined by

$$\Lambda\gamma^i = e^i \tag{2.2.16}$$

So we have

$$u_i = (w^i, z_d)_H. \tag{2.2.17}$$

It results that we can write

$$u = [\, u_1 \cdots u_N\,] = \left[(w^1, z_d)_H \cdots (w^N, z_d)_H\right] \tag{2.2.18}$$

Comparing (2.2.9) and (2.2.18), we identify the generalized inverse W and

$$W = \left[w^1 \cdots w^N \right]^T \tag{2.2.19}$$

At first we have to determine B. For i from 1 to N, let ψ_i denote the function defined by (2.1.1) and which is the solution of the state equation (2.1.1) when all the parameters v_j of v, but the i-th, are zero, and $v_i = 1$. We have seen that the contribution of the i-th source to the observation z of state y is

$$Be^i = C\Psi_i \tag{2.2.20}$$

Here from the mathematical point of view, the N_1 actual sources of pollution and the N_2 meshes play the same role.

2.2.2 Direct method

Algorithm 2.2.1

1. Compute $Be^i \in H$, $1 \leq i \leq N$,

2. Calculate the elements of matrix Λ:

$$\Lambda_i^j = \int_{t_0}^{t_1} C\Psi_i(t) C\Psi_j(t) \, dt \tag{2.2.21}$$

for each parameter needing identification, for example, the n-th one.

3. We define γ^n by

$$\Lambda\gamma^n = e^n \tag{2.2.22}$$

4. We define w^n by

$$w^n = B\gamma^n \tag{2.2 23}$$

Remark 2.2.1

Algorithm 2.2.1 does not necessitate calculating an adjoint state.

Remark 2.2.2

The direct method needs the values of the N functions ψ_i at the M observation points x_k. We therefore have to solve the state equations N times. If we have at our disposal a formula giving explicitly the solution of these equations for each (x, t), it may be adequate to use this formula at each of the observation points x_k. That is the case in particular when the domain Ω has a simple geometry, for example, if it is a segment, a square quadrilateral, or a cube. The case of a square will be developped in chapter 3. The direct method can, of course, be implemented for various geometries and modellings, by using, for example, a finite element method. But its drawback is the large number N of times the state equation has to be solved. We have $N = N_1 + N_2$, where the number N_1, of sources, is small, for example, $N_1 = 1$ or 2, whereas the number N_2, of finite elements is large, for example $N_2 = 100\ 000$. A remedy for this problem is to use a partition of Ω by only a few super-elements Ω_j, or to use the indirect method.

Remark 2.2.3

We suppose that the sensors continuously monitor the concentrations at each measurement point and that consequently the observations at each time $t \in\ ]\ 0,T[$ are available. However the method applies equally well to the situation of discrete measurements, provided the sampling rate is large enough or the period of observation is long enough. Anyway in order to perform the numerical simulations we discretize with respect to time. The ideal would be many data on a long interval. But it is not the general case.

2.2.3 Indirect method

When N is large, a method avoiding calculation of Λ would be most welcome. Concerning the n-th source, solving $\Lambda\gamma^n = e^n$ is equivalent to minimizing the cost function

$$J(\gamma) = \frac{1}{2}(\Lambda\gamma, \gamma)_V - (e^n, \gamma)_{V^4} \qquad (2.2.24)$$

which can be rewritten as

$$J(\gamma) = \frac{1}{2}(B\gamma, B\gamma)_H - (e^n, \gamma)_V \qquad (2.2.25)$$

For a given control $\gamma = (\alpha, \beta)$ let $\rho(\gamma)$ denote the solution of the linear problem

$$\begin{cases} \rho' + A\rho = \sum_{i=1}^{i=N_1} \alpha_i s_i(t)\delta(x - a_i) \\ \rho(x,0) = \sum_{j=1}^{j=N_2} \beta_j \chi_j(x), \, x \in \Omega \\ \dfrac{\partial \rho}{\partial n} = 0 \end{cases} \qquad (2.2.26)$$

(linear in the sense that the observation linearly depends upon the parameters γ). This observation is the element $\{\rho(x_k, t; \gamma)\}_{1 \le k \le M}$ of H where the state $\rho(\gamma)$ is the solution of the system of equations

The components of the control vector γ are the N_1 parameters α_i and the N_2 parameters β_j:

$$\gamma = (\alpha, \beta), \alpha = \{\alpha_i\}, 1 \le i \le N_1, \beta = \{\beta_j\}, 1 \le j \le N_2 \qquad (2.2.27)$$

$$\gamma = (\alpha, \beta) \in R^{N_1} \times R^{N_2} = R^N \qquad (2.2.28)$$

$$J(\gamma) = \frac{1}{2}|C\rho(\gamma)|_H^2 - (e^n, \gamma)_V = \frac{1}{2}\sum_{k=1}^{k=M}\int_0^T \rho^2(x_k, t; \gamma)dt - (e^n, \gamma)_V \qquad (2.2.29)$$

Let us consider the linear-quadratic optimal control problem where γ is the control, $\rho(\gamma)$ is the corresponding state, and (2.2.29) defines the cost function. Then it is equivalent to minimize (2.2.25) without any constraint on γ or to minimize (2.2.29) with the constraints (2.2.23).

Algorithm 2.2.2

1. Solve the optimal control problem: minimize (2.2.25) under the constraints (2.2.26).

2. Once the optimal control γ^n is known, the sentinel w^n is given by

$$w'' = B\gamma'' = C\rho\left(\gamma''\right).$$

(2.2.31)

2.2.4 Gradient methods

Gradient methods are based upon the fact that, if we know the vector of controls $\gamma = (\alpha, \beta)$, then the cost function $J(\gamma)$ and its gradient $J'(\gamma)$ are obtained by solving the following two systems of equations for, respectively, the state r and the adjoint state q :

$$
\begin{cases}
\rho' + A\rho = \displaystyle\sum_{i=1}^{i=N_1} \alpha_i s_i(t)\delta(x - a_i) \\[2ex]
\rho(x,0) = \displaystyle\sum_{j=1}^{j=N} \beta_j \chi_j(x) \; on \; \Omega \\[2ex]
\dfrac{\partial p}{\partial n} = 0 \; on \; \Sigma
\end{cases}
$$

(2.2.32)

$$
\begin{cases}
-q' + A^* q = C^* w = \displaystyle\sum_{k=1}^{k=M} w_k(t)\delta(x - x_k) \\[2ex]
q(x,T) = 0 \; on \; \Omega \\[2ex]
\dfrac{\partial q}{\partial n} = 0 \; on \; \Sigma
\end{cases}
$$

(2.2.33)

with

$$w_k(t) = \rho\left(x_k, t\right)$$

(2.2.34)

We have employed the conjugate gradient method, but only as an example among many possible others. Other gradient methods may be better, but the conjugate gradient method performs well. In particular, it converges in a few iterations. This algorithm is well-known.for the minimization of the cost function :

$$J(\gamma) = \frac{1}{2}(\Lambda\gamma, \gamma)_V - (b, \gamma)_V$$

(2.2.35)

Let us make more precise the details of the application of this algorithm to our case, where $b = e^n$ and where we do not have an explicit expression of the matrix Λ. In order to obtain the value of the cost function $J(\gamma)$ and its gradient $J'(\gamma)$ we use a function called Funct for example, whose input parameters are [b, g, flag] and output parameters are [J, g,w] (J = cost function $J(\gamma)$, g = gradient. $J'(\gamma)$)) b second member of equation $\Lambda\gamma = b$, γ solution of this equation.

$$[J, g, w] = \text{Funct} (\beta, \gamma, \text{flag}) \tag{2.2.36}$$

According to whether Flag = 0 or 1 we only calculate J and w or we calculate J, w, and also $g = J'(\gamma)$.

To calculate w we first solve the pseudo-state equations (2.2.32), which provide the M components of w

$w_k(t) = \rho(x_k,t)$ (that element of H which hopefully converges

towards w^n) and the corresponding value of the cost function:

$$J(\gamma) = \frac{1}{2} C\rho \big|^2_H - (b, \gamma)_V \tag{2.2.37}$$

or, equivalently,

$$J(\gamma) = \frac{1}{2} B\gamma \big|^2_H - (e^n, \gamma)_V$$

If flag = 0 we stop there.

If flag = 1 we solve the "adjoint" system (2.2.33), and we obtain the gradient vector g, of components

$$\begin{cases} \left\{ \int_0^T s_i(t)q(a_i,t)dt - \delta_i^n \right\}_{1 \le i \le N_1} \\ \left\{ \int_{\Omega_j} q(x,0)dx \right\}_{1 \le j \le N_2} \end{cases}$$

$$\begin{cases} \left\{ \int_0^T s_i(t)q(a_i,t)dt - \delta_i^n \right\}_{1 \le i \le N_1} \\ \left\{ \int_{\Omega_j} q(x,0)dx \right\}_{1 \le j \le N_2} \end{cases} \tag{2.2.38}$$

2.2.5 Conjugate Gradient

.initialisation: x = 0. comments: {for example} ; {x = γ = (α, β)}.

[w, J, g] = Funct (b, γ, 1) comments: instead of g = $\Lambda\gamma - b$ and
$J = \frac{1}{2}(\Lambda\gamma, \gamma) - (b, \gamma)$

.if norm (g) $\leq \varepsilon$ end, Otherwise: g_2 = (g, g) ; d = -g ;

.end of initialization

.iterations: While norm (g) > ε,

[w, J] = Funct(0, d, 0)

$\mu = d^T g/(2 \times J)$,

$\gamma = \gamma + \mu$ d,

[w, J, g] = Funct (b, γ, 1). comments: instead of g = $\Lambda \gamma$ - b and
$J = \frac{1}{2}(\Lambda\gamma, \gamma) - (b, \gamma)$

g_1 = g_2 ; g_2 = (g, g) ; β = g_2/g_1 ; d = -g+β d ;

.end of iterations.

2.2.6 Cost and gradient

The interesting point is that it is not necessary, to obtain $(\Lambda d, d)_v$ = (Bg, Bg) to know the matrix Λ. When g is known, various methods (finite differences, finite elements, explicit expression, *etc.*) give the solution ρ of (2.2.33) and therefore p. Let us denote γ^n the vector minimizing J(g), (n = number of the parameter and of the associated sentinel). The n-th sentinel is:

$$w^n = B\gamma^n = C\rho(\gamma^n) \tag{2.2.39}$$

This sentinel w^n is the observation of the state $\rho(\gamma^n)$.

Remark 2.4

We see the interest of an indirect method (Algorithm 1.3 with for example the conjugate gradient to minimize $J(\gamma)$: each iteration of the conjugate gradient method only necessitates a few (here three) calculations of state or adjoint state. If (as it is the case in the examples given Section 3) a good convergence is obtained as early as at the third iteration of the conjugate gradient method, the determination of a sentinel only needs nine calculations of the state or its co-state.

This is to be compared to the N calculations of a state ψ_i for the so-called direct method!

Remark 2.5

The reason why $N = N_1 + N_2$ is large is that N2, which is the number of finite elements, is large. In fact there are too many missing degrees of freedom and the experience shows that in general N_2 is much too large but is it is possible to achieve a quasi insensitivity to initial conditions with a coarser partition of W. Moreover the initial conditions contribute to the solution of the parabolic equation by terms that tend exponentially to 0. The larger the time interval, the smaller the influence of these initial conditions on the behavior of the.system. As the time-interval $] 0, T[$ increases the state and therefore the observation $y(x_k, t)$ on this time-interval naturely become less and less sensitive to initial conditions.

2.2.7 Parameter estimation

For a given observation z_d, the n-th parameter u_n is given by the inner product in H

$$u_n = (w^n, z_d)_H$$

$$(2.2.40)$$

Remark 2.2.6

An advantage of the sentinels method is the possibility of pre-calculating a sentinel, then using it repeatedly for a fast treatment of incoming new data.

Another advantage of the sentinels methodis that as a by-product it gives an indication of how much each parameter value can be perturbated by noise. We have:

$$u_i = (w^i, z_d)_H \Rightarrow |u_i|_{\Re} \le |w^i|_H |z_d|_H \qquad (2.2.41)$$

so that $|w^i|_H$ gives this information.

Now we consider the family of all sentinels those attached to pollution parameters λ_i, $1 \le i \le N_1$, and those attached to missing parameters τ_j, $1 \le j \le N_2$, and we let

$$W = [w^1, w^2, \ldots, w^N]^T \in H^N \qquad (2.2.42)$$

W can be identified with the generalized inverse of B. There are N sentinels w^n, w^n being the sentinel associated with parameter v^n. w^n itself is defined by M functions of L^2 (]0,T[)

2.3 Numerical experiment

The finite elements mesh:

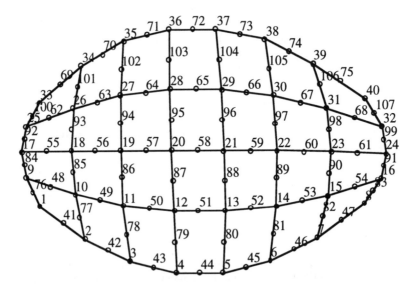

Figure 2.3

The domain and its mesh by eight-nodes quadrilaterals. We can see that there are 107 nodes, and, consequently, since the boundary conditions are of null flux type, there are also 107 degrees of freedom.

2.3.1 Representation of the source and of observation points

S = Source   O = Observatory

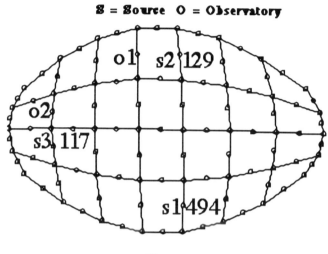

Figure 2.4

Source points (s) together with the actual intensity (494, 129, and 117). Observation points (o1, and o2).

2.3.2 Operation mode

Here is now the procedure that enables us to test the aptitude of the sentinels to identify the coefficients λ_n. Once the sentinel w^n of the n-th source has been calculated :

1. We choose parameter values (λ, τ) for which we solve the state equation, thus obtaining in particular the solution at measurement points. We shall say that these parameters values are exact.

2. Next we add noise to the values of the solution at measurement points with the following convention, when we say that the $\dfrac{noise}{signal}$ ratio is x%, we mean that the observed values of y are randomly uniformly distributed according to a uniform law in the interval

$$\left((1-\frac{x}{100})\times y_{exact},(1+\frac{x}{100})\times y_{exact}\right).$$

For example, when the $\dfrac{noise}{signal}$ ratio is 20%,

$$0.8 \times y_{exact} \leq y_{measured} \leq 1.2 \times y_{exact}.$$

In that case we will talk of a noise of 20%. It is important that the estimated parameter values continue to be close to the exact ones even though such a noise perturbs the observation.

3.Then we calculate the inner-product in H of the sentinel w^n of the n-th source, and of this more or less noisy observation z, and we obtain an estimate of the intensity λ_n of this source.

The following table gives, for different values of noise, the values λ_n estimated for some noise levels.

Table 2.1

	λ_n exact	Noise 20%	Noise 10%	Noise 5%
source 1	117.0000	174.5185	130.2557	131.5803
source 2	129.0000	152.1114	131.1738	130.5569
source 3	494.0000	416.5176	495.1532	488.0900

An examination of Figure 2.4 enables understanding why the approximations of λ_2 and λ_3 are better than those of λ_1. Each source 2 and 3 is close to a point of observation.

2.4 Adjoint state

We present now another approach to the sentinels.

2.4.1 Adjoint state

We begin by giving this definition of a sentinel:

Definition 2.4.1

Let z be the observation of the state y of the system $z = y/(\omega \times]0,T[)$ in
$H = \left(L^2(]0,T[\times R^M)\right)$

A sentinel is a linear functional S that associates with the observation z the number

$$S(z) = \sum_{k=1}^{k=M} \int_0^T w_k(t) z_k(t) dt \qquad (2.4.1)$$

where

$$z_k(t) = y(x_k,t), \quad x_k \text{ observation point. } Ay = -\Delta y + \sigma y, \quad A = -\Delta + \sigma,$$

Thus a sentinel is defined by M functions $w_k: t \to w_k(t)$ of $L^2(]0,T[)$ (M number of points of observation in the observatory). We identify S with $w = \{w_k\}_{1 \le k \le M} = \{w_1, \cdots, w_M\}$, element of $H = \left(L^2(]0,T[\times R^M)\right)$.

Remark 2.4.1

Here we suppose that the observation z is exact. Thus $z = Cy$ and
$S(z) = (w,z)_H = (w,Cy)_H$
Let us define the adjoint state q:

$$\begin{cases} -q' + A^* q = \sum_{k=1}^{k=M} w_k(t)\delta(x - x_k) \text{ on } Q = \Omega \times]0,T[\\ q(x,T) = 0 \text{ on } \Omega \end{cases} \qquad (2.4.2)$$

The "zero flux" boundary condition on $\Sigma = \Gamma \times]0, T[$ is implicitly included in the formulation (2.4.2).

We have successively:

$$S(z) = \int_Q \left\{ \sum_{k=1}^{k=M} w_k(t)\delta(x - x_k) \right\} y(x,t)dxdt$$

$$S(z) = \int_Q (-q' + A^* q)ydxdt = \int_Q q(y' + Ay)dxdt + \int_\Omega q(x,0)y(x,0)dx$$

$$S(z) = \int_Q q \sum_{i=1}^{i=N_1} \lambda_i s_i(t)\delta(x - a_i)dxdt + \int_\Omega q(x,0) \sum_{j=1}^{j=N_2} \tau_j \chi_j(x)dx$$

$$S(z) = \sum_{i=1}^{i=N_1} \lambda_i \int_0^T s_i(t)q(a_i,t)dt + \sum_{j=1}^{j=N_2} \tau_j \int_{\Omega_j} q(x,0)dx$$

$$S(z) = \sum_{i=1}^{i=N_1} c_i\lambda_i + \sum_{j=1}^{j=N_2} d_j\tau_j \tag{2.4.3}$$

provided

$$\begin{cases} \int_0^T s_i(t)q(a_i,t)dt = c_i & 1 \le i \le N_1 \\ \int_{\Omega_j} (q(x,0))dx = d_j & 1 \le j \le N_2 \end{cases} \tag{2.4.4}$$

whence

Proposition 2.4.1

The function w being given in $H = (L^2(]0,T[\times R^M)$, the value S(z) of the sentinel defined by this w in (2.4.1) satisfies (2.4.3) if and only if the solution q of (2.4.2) satisfies the constraints (2.4.4).

Remark 2.4.2

In particular for all the coefficients in c and d equal to 0, excepted $c_n = 1$ we have $S(z) = \lambda_n$.

For all the coefficients in c equal to 0, and the coefficients in d equal to $d_j = \int_{\Omega_j} 1 dx$, $1 \le j \le N_2$, we have. $S(z) = \sum\limits_{j=1}^{j=N_2} \tau_j \int_{\Omega_j} 1 dx$, the total initial amount of contaminant.

Proposition   2.4.2

The coefficients $c = \{c_i\}_{1 \le i \le N_1}$ and $d = \{d_j\}_{1 \le j \le N_2}$ being given, we have

$$S(z) = \sum_{i=1}^{i=N_1} c_i \lambda_i + \sum_{j=1}^{j=N_2} d_j \tau_j$$

if and only if the function q which corresponds to w by (2.4.2) satisfies the conditions (2.4.4).

Thus we are faced with an exact controllability problem:

The adjoint state q being governed by equations (2.4.2) find functions t $\in w_k(t)$, k = 1, ..., M controlling it (in retrograde time) in such a way that the constraints (2.4.4) are satisfied.

To solve this problem we make the following fundamental remark:

Remark   2.4.3

The conditions (2.4.2), together with $w_k(t) = r(x_k, t)$, and (2.4.4) are the definition of the adjoint stateand the optimality conditions for the problem of control defined by the following state equation, initial condition, vector of control, and cost function:

$$
\begin{cases}
\rho' + A\rho = \sum\limits_{i=1}^{i=N_1} \alpha_i s_i(t)\delta(x - a_i) \\[2mm]
\rho(x,0) = \sum\limits_{j=1}^{j=N_2} \beta_j \chi_j(x) \\[2mm]
\gamma = (\alpha, \beta), \ \alpha = \{\alpha_i\}_{1 \le i \le N_1}, \beta = \{\alpha_j\}_{1 \le j \le N_2} \\[2mm]
J(\alpha, \beta) = \frac{1}{2} \sum\limits_{k=1}^{k=M} \int_0^T \rho^2(x_k, t)dt - \sum\limits_{i=1}^{N_1} c_i \alpha_i - \sum\limits_{j=1}^{N_2} d_j \beta_j
\end{cases}
\qquad (2.4.5)
$$

We have the following existence result.

Proposition   2.4.3

The vectors $c = \{c_i\}_{1 \leq i \leq N_1}$ *and* $d = \{d_j\}_{1 \leq j \leq N_2}$ *being given, there exists a sentinel* $\{w_k(t) = \rho(x_k, t)\}_{1 \leq k \leq M}$, *such that (2.4.3) holds.*

proof

It results immediately from the linearity of the operator $B: \gamma = (\alpha, \beta) \rightarrow C\rho$ that

$$\rho(x_k, t; \gamma) = \sum_{i=1}^{i=N} \gamma_i \Psi_i(x_k, t) \tag{2.4.6}$$

where Ψ_i is the solution of the state equations in (2.4.5) when all the coefficients in γ are zero, except $\gamma_i = 1$. Consequently

$$J(\gamma) = \frac{1}{2}(\Lambda\gamma, \gamma) - (b, \gamma) \tag{2.4.7}$$

where $b = \begin{vmatrix} c \\ d \end{vmatrix}$ and Λ is the symmetric $N \times N$ matrix defined in (2.2.7), positive definite provided the Ψ_i / w are linearly independent, of general term.

$$\Lambda_i^j = \sum_{k=1}^{k=M} \int_0^T \Psi_i(x_k, t)\Psi_j(x_k, t)dt = \left(C\Psi_i, C\Psi_j\right)_H. \tag{2.4.8}$$

We thus are in the most comfortable situation of numerical analysis, in which the problem is to minimize a quadratic form with a symmetric and positive definite matrix Λ. The solution of this problem is also solution of the linear equation

$$\Lambda\gamma = b. \tag{2.4.9}$$

Knowing γ, we get

$$w_k(t) = \rho\left(x_k, t; \gamma\right) \ 1 \leq k \leq M \tag{2.4.10}$$

In the case where we already know the functions Y_j, we have

$$w_k(t) = \rho(x_k, t; \gamma) = \sum_{i=1}^{i=N} \gamma_i \Psi_i(x_k, t) \tag{2.4.11}$$

Instead we have to solve for ρ the state equations.

Remark 2.4.4

The gradient of J is the N-dimensional vector whose elements are

$$\left\{ \int_0^T s_i(t) q(a_i, t) dt - c_i \right\}_{1 \le i \le N_1}, \left\{ \int_{\Omega_j} q(s, 0) dx - d_j \right\}_{1 \le j \le N_2} \tag{2.4.12}$$

It results that, by a suitable choice of the coefficients c_i and d_j, one can get information about the vectors of parameters l and t. We now consider some examples.

2.4.2 Intensity of a source of pollution

When all the coefficients d_j are zero, and also all the coefficients c_i, excepted one, say $c_n = 1$, then the value of the sentinel is λ_n.

Proposition 2.4.4

There exists a function w in $L^2(]0,T[;R^M)$ such that the solution q of (2.4.2) satisfies

$$\begin{cases} \int_0^T s_i(t) q(a_i, t) dt = \delta_i^n & 1 \le i, \, n \le N_1 \\ \int_{\Omega_j} q(x, 0) dx = 0 & 1 \le j \le N_2 \end{cases} \tag{2.4.13}$$

The corresponding value S(z) of the sentinel defined in (2.4.1) by this function w is

$$S(z) = \lambda_n \tag{2.4.14}$$

$$\begin{cases} c_i = \delta_i^n \ \ for \ 1 \le i \le N_1 \\ d_j = 0 \ \ 1 \le j \le N_2 \end{cases} \tag{2.4.15}$$

$$\sum_{k=1}^{k=M} \int_0^T w_k(t) z_k(t) dt = \lambda_n \tag{2.4.16}$$

2.4.3 Local intensity of initial pollution

When all the coefficients c_i are zero, together with all the coefficients d_j, excepted one, say $d_n = 1$, then the value $S(z)$ of the sentinel is

$$S(z) = \tau_n \tag{2.4.17}$$

Proposition 2.4.5

There exists a function w in $L^2(]0,T[;R^M)$ such that the solution q of (2.4.2) satisfies

$$\int_0^T s_i(t) q(a_i, t) dt = 0 \qquad 1 \le i, \ n \le N_1 \ \text{and}$$

$$(q(0), \chi_j)_{L^2(\Omega)} = \delta_j^n, \ 0 \le j \le N \tag{2.4.18}$$

Suppose now that

$$c_i = \int_0^T s_i(t) dt \ \ 1 \le i \le N_1$$

$$d_j = 0 \ \forall j \ \ 1 \le j \le N_2. \tag{2.4.20}$$

$$\int_0^T s_i(t) q(a_i, t) dt = \int_0^T s_i(t) dt, 1 \le i \le N_1$$

$$\int_{\Omega_j} q(x,0) dx = 0 \ \ 1 \le j \le N_2 \tag{2.4.21}$$

The value S(z) of the sentinel defined in (2.4.1) by this function w is

$$S(z) = \sum_{i=1}^{i=N_1} c_i \lambda_i = \sum_{i=1}^{i=N_1} \lambda_i \int_0^T s_i(t) dt \qquad (2.4.22)$$

The value of such a sentinel is the total amount of pollutant discharged by the N_1 sources of pollution during the interval of time (0,T). Hence we are required to choose the functions w_k such that the solution q of (2.4.2) satisfy (2.4.21).

Proposition 2.4.6

There exists a function $w \in L^2(0,T;R^M)$ such that the solution q of (2.4.2) satisfies (2.4.21). The value S(z) of the sentinel defined in (2.4.1) by this function w is (2.4.22).

Table 2.2 comparing the exact and identified total discharge

Noise	0	20%	100%
Exact discharge	53.0	53.0	53.0
Estimated discharge	53.1	53.1	53.0

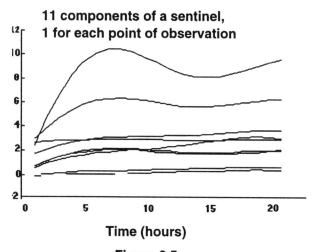

11 components of a sentinel, 1 for each point of observation

Time (hours)

Figure 2.5

2.4.5 Total initial charge of pollutant

Suppose now that

$$c_i = 0 \ 1 \le i \le N_1 \text{ and } d_j = \int_{\Omega_j} 1 dx \ \ 1 \le j \le N_2 \tag{2.4.23}$$

$$\int_0^T s_i(t)q(a_i,t)dt = 0, \ 1 \le i \le N_1 , \tag{2.4.24}$$

and

$$\int_{\Omega_j} q(x,0)dx = \int_{\Omega_j} 1 dx \ \ 1 \le j \le N_2 \tag{2.4.25}$$

The value S(z) of the sentinel defined in (2.4.1) by this fonction w is

$$S(z) = \sum_{j=1}^{j=N_2} \int_{\Omega_j} 1 dx \tau_j \tag{2.4.26}$$

The value of such a sentinel is the sum of pollutant initial charges in the N_2 finite element meshes. It is the total amount of pollutant in the domain Ω at time zero. Hence we are required to find the functions w_k such that the solution q of (2.4.2) satisfies (2.4.21) and (2.4.22).

Proposition 2.4.7

There exists a function w ∈ $L^2(]0,T[; R^M)$ such that the solution q of (2.4.2) satisfies (2.4.24).and (2.4.25). The value S(z) of the sentinel defined in (2.4.1) by this fonction w is (2.4.26), the total amount of pollutant present in the domain Ω. at initial time.

Thus our goal is to determine this function w. To acheive this goal we proceed as explained in detail in the next section.

2.5 Numerical details

Let us examine how to design a program that calculates this total initial pollution. We have the equations of state

$$\begin{cases} y' + Ay = \sum_{i=1}^{N_1} \lambda_i s_i(t)\delta(x - a_i) \text{ in } Q = \Omega \times]0, T[\\ \dfrac{\partial y}{\partial n} = 0 \text{ in } \Sigma = \Gamma \times]0, T[\\ y(x,0) = \sum_{j=1}^{N_2} \tau_j \chi_j(x) \text{ in } \Omega \end{cases},$$

and the data provided by the observation of this state

$$z_k : t \to z_k(t) \equiv y\left(x_k, t; v\right) \ 1 \le k \le M \tag{2.5.1}$$

The indirect method to obtain the sentinel w is the conjugate gradient method applied to the minimization of the cost function

$$J(\alpha, \beta) = \frac{1}{2} \sum_{k=1}^{k=M} \int_0^T \rho^2(x_k, t)dt - \sum_{i=1}^{N_1} c_i \alpha_i - \sum_{j=1}^{N_2} d_j \beta_j$$

where ρ is the solution of

$$\begin{cases} \rho' + A\rho = \sum_{\substack{i=1}}^{i=N_1} \alpha_i s_i(t)\delta(x - a_i) \\ \rho(x,0) = \sum_{j=1}^{j=N_2} \beta_j \chi_j(x) \end{cases} \tag{2.5.2}$$

and, in this specific example,

$$c_j = 0, \ 1 \le i \le N_1 \text{ and } d_j = \int_{\Omega_j} 1 dx, \ 1 \le j \le N_2.$$

In order for

$$S(z) = \sum_{i=1}^{i=N_1} c_i \lambda_i + \sum_{j=1}^{j=N_2} d_j \tau_j \tag{2.5.3}$$

to hold we require the parameters $\gamma = (\alpha, \beta)$, $\alpha = \{\alpha_i\}_{1 \le i \le N_1}$, $\beta = \{\alpha_j\}_{1 \le j \le N_2}$ to be such that

$$\begin{cases} \int_0^T s_i(t)q(a_i,t)dt = c_i, \ 1 \le i \le N_1 \\ \int_{\Omega_j}(q(x,0))dx = d_j, \ 1 \le j \le N_2 \end{cases},$$

q being the solution of the equations

$$\begin{cases} -q' + A^*q = \sum_{k=1}^{k=M} \rho(x_k,t)\delta(x-x_k) \text{ on } Q = \Omega \times] \, 0,T[\\ q(x,T) = 0 \ on \ \Omega \end{cases}$$

It is equivalent to minimize (2.5.1) under the constraints (2.5.2) or to minimize

$$J(\gamma) = \frac{1}{2}|B\gamma|_H^2 - \sum_{j=1}^{N_2} d_j \beta_j \tag{2.5.4}$$

without any constraint. Recall that we know the coefficients

$$d_j = \int_{\Omega_j} 1 \ dx, \ 1 \le j \le N_2$$

Once the minimizer $\tilde{\gamma}$ has been found then the sentinel is given by $w = B\tilde{\gamma} = C\rho(\tilde{\gamma})$ and once the sentinel has been found the total initial pollution is given by $f = (w,z)_H$

2.6. Hum method

HUM means Hilbert uniqueness method, a method generalizing to the case of an infinite-dimensional space V what we did for an n-dimensional space V. Let us recall what we have done: we are interested in a sentinel w such that

$$S(z) = \sum_{i=1}^{i=N_1} c_i \lambda_i + \sum_{j=1}^{j=N_2} d_j \tau_j \tag{2.6.1}$$

this sentinel w is given by

$$w = C\rho \tag{2.6.2}$$

where ρ is the solution of

$$\begin{cases} \rho' + A\rho = \sum_{\substack{i=1 \\ j=N_2}}^{i=N_1} \alpha_i s_i(t)\delta(x - a_i) \\ \rho(x,0) = \sum_{j=1} \beta_j \chi_j(x) \end{cases} \tag{2.6.3}$$

and the parameters

$$\gamma = (\alpha, \beta), \quad \alpha = \{\alpha_i\}_{1 \le i \le N_1}, \beta = \{\beta_j\}_{1 \le j \le N_2} \tag{2.6.4}$$

are such that, if q is the solution of

$$\begin{cases} -q' + A^* q = \sum_{k=1}^{k=M} \rho(x_k, t)\delta(x - x_k) \text{ on } Q = \Omega \times]0, T[\\ q(x,T) = 0 \text{ on } \Omega \end{cases}, \tag{2.6.5}$$

then the following constraints are fulfilled:

$$\begin{cases} \int_0^T s_i(t)q(a_i,t)dt = c_i \ 1 \le i \le N_1 \\ \int_{\Omega_j} (q(x,0)dx = d_j \ 1 \le j \le N_2 \end{cases}. \tag{2.6.6}$$

therefore, once we know ρ, w is easily calculated. Equations (2.6.3) define the operator B: $R^N \rightarrow H$

$\gamma \rightarrow \begin{array}{c} w = C\rho \end{array}$. Equations (2.6.5) and (2.6.6) define its adjoint $B^*: H \rightarrow R^N$

$$w \rightarrow \left\{ \left\{ \int_0^T s_i(t)q(a_i,t)dt \right\}_{1 \le i \le N_1}, \left\{ \int_{\Omega_j} (q(x,0)dx \right\}_{1 \le j \le N_2} \right\}$$

the problem is to solve for γ the equation

$$\Lambda\gamma = b \tag{2.6.7}$$

where

$$\Lambda = B^* B \text{ and } b = \{c,d\}, \ c = \{c_i\}_{1 \le i \le N_1}, d = \{d_j\}_{1 \le j \le N_2}. \tag{2.6.8}$$

We recognize here the formulation already encountered several times, to which we can apply either the direct or the indirect method.

2.7 Time and space discretization

Since the solutions of our PDEs are functions of time and space, let us explain the numerical schemes we adopt with respect to time and space.

2.7.1 Time discretization

Consider the system

$$\rho' + A\rho = \sum_{i=1}^{i=N_1} \alpha_i s_i(t)\delta(x - a_i) \tag{2.7.1}$$

$$\rho(x,0) = \sum_{j=1}^{j=N_2} \beta_j \chi_j(x) \tag{2.7.2}$$

We partition the time interval (0, T) into N_3 elementary sub-intervals of length $\Delta t = \dfrac{T}{N_3}$ and we call $\mathbf{r}^n$ the approximation of r(x, nΔt) defined for n = 1, 2, ... , N_3 by

$$\frac{\mathbf{r}^n - \mathbf{r}^{n-1}}{\Delta t} + A\mathbf{r}^n = \sum_{i=1}^{i=N_1} \alpha_i s_i \delta(x - a_i) \tag{2.7.3}$$

$$\mathbf{r}^0(x) = \sum_{j=1}^{j=N_2} \beta_j \chi_j(x) \tag{2.7.4}$$

To this time discretization corresponds the discretized cost function

$$J(\alpha,\beta) = \frac{\Delta t}{2}(w_k^n)^* w_k^n \tag{2.7.5}$$

where

$$w_k^n = \mathbf{r}^n(x_k).$$

2.7.2 Space discretization

We call $\mathbf{H}_i$ the i-th nodal function affected to node i, $\mathbf{H}(x)$ the row vector of the values of thesenodal functions at point x. $H_i(x)$ is non-zero only if the i-th node is adjacent to the element to which x belongs. For example the approximation of function r can be written $\hat{\mathbf{r}}(x) = H(x)\mathbf{r}$ where $\mathbf{r}$ is the column vector of all the nodal values $\hat{\mathbf{r}} = \mathbf{Hr}$. Similarly for a test vector $\mathbf{v}$, $\hat{\mathbf{v}} = \mathbf{Hv}$.

The time and space discretized state equation then can be written as

$$\left(K + \frac{1}{\Delta t}M\right)r^n = \sum_{i=1}^{i=N_1} a_i s_i^n b_i + \frac{1}{\Delta t}Mr^{n-1}$$ (2.7.6)

where K and M respectively are the stiffness matrix, the mass matrix and

$$b_i = \sum_j H_j^T(a_i)$$ (2.7.7)

is the sum of the nodal functions H_j which are non-zero on the element to which a_i belongs, i.e. related to the nodes adjacent to this element.

The discretized initial condition is

$$r^0(x) = \sum_{j=1}^{j=N_2} \beta_j \chi_j(x)$$

The discretized cost function is

$$J(\alpha,\beta) = \frac{\Delta t}{2} \sum_{k=1}^{M} \sum_{n=1}^{N_3} (r^n)^T H(x_k)^T H(x_k) r^n - \sum_{i=1}^{N_1} c_i \alpha_i - \sum_{j=1}^{N_2} d_j \beta_j$$

$$J(\alpha,\beta) = \frac{\Delta t}{2} \sum_{k=1}^{M} \sum_{n=1}^{N_3} (w_k^n)^T w_k^n - \sum_{i=1}^{N_1} c_i \alpha_i - \sum_{j=1}^{N_2} d_j \beta_j$$ (2.7.8)

The discretized observation is

$$\mathbf{w}_k^n = \mathbf{H}(x_k)\mathbf{r}^n \tag{2.7.9}$$

observation at point x_k and time n Dt.

2.7.2 Adjoint of the discretized problem

It is well-known that, in order to solve an optimal control problem, one has to be careful to use the adjoint of the discretized problem rather than the discretized adjoint. We recall below how it can be found using a Lagrangian. The Lagrangian L is the function of the independent variables $\{\alpha, \beta, r, q\}$ defined by

$$
L(\alpha,\beta,r,q) = \frac{\Delta t}{2} \sum_{k=1}^{M} \sum_{n=1}^{N_3} \mathbf{w}_k^{n^T} \mathbf{w}_k^n
$$
$$
-\Delta t \sum_{n=1}^{n=N_3} \mathbf{q}^n \left(\left(K + \frac{1}{\Delta t}M\right)\mathbf{r}^n - \sum_{i=1}^{i=N_1} \alpha_i s_i^n b_i - \left(\frac{1}{\Delta t}M\right)\mathbf{r}^{n-1} \right)
$$

where w is given by (1.25), (.,.) inner product in R^n, and n is the number of degrees of freedom, here equal to the number of nodes. The gradient of J is given by

$$
\frac{\partial J}{\partial \alpha_i} = \frac{\partial L}{\partial \alpha_i} = \Delta t \sum_n s_i^n \left(\mathbf{q}^n, b_i \right) - c_i, 1 \le i \le N_1 \tag{2.7.10}
$$

$$
\frac{\partial J}{\partial \beta_j} = \frac{\partial L}{\partial \beta_j} = \left(M\chi_j, \mathbf{q}^1 \right) - d_j, 1 \le j \le N_2 \tag{2.7.11}
$$

provided the adjoint state $\mathbf{q}$ satisfies the condition

$$
\frac{\partial L}{\partial r^n} \hat{r}^n = 0 \quad \forall \hat{r}^n \in R^n
$$

$$
\frac{\partial L}{\partial r^n} \hat{r}^n = \Delta t \sum_{k=1}^{M} \sum_{n=1}^{N_3} \mathbf{w}_k^{n^T} \hat{\mathbf{w}}_k^n - \Delta t \sum_{n=1}^{n=N_3} \left(\mathbf{q}^n, \left(K + \frac{1}{\Delta t}M\right)\hat{r}^n - \frac{1}{\Delta t}\left(\mathbf{q}^{n+1}, M\hat{r}^n\right) \right)
$$

$$\sum_k w_k^n \hat{w}_k^n = \sum_k w_k^n H(x_k) \hat{\rho}_k^n = \sum_k w_k^n \left(b_k \hat{\rho}^n \right)$$

where

$$H(x_k) = \left\{ H_1(x_k), \cdots, H_{ndof}(x_k) \right\} \text{ and } b_k = H^T(x_k).$$

Note that b_k is a column vector $\in R^{ndof}$. Its only eventual non zero components are those corresponding to nodes j adjacent to the element where is the observation point x_k. At last the adjoint state q^n is defined by

$$(K + \frac{1}{\Delta t} M) q^n = \sum_{k=1}^{k=M} w_k^n b_k + \frac{1}{\Delta t} M q^{n+1}, 1 \le n \le N_3 \qquad (2.7.13)$$

$$q^{N_3+1} = 0 \qquad (2.7.14)$$

2.8 Optimal emplacement of sensors

To illustrate this aspect of the method of sentinels, let us consider the configuration of two sources and one observatory represented Figure 2.7.

Both sources have the same amplitude, 1, and the sensor is closer to source 2 than to source 1. Therefore we expect source 2 to be better identified than source 1.

The following two tables show, on each line, from left to right, the conjugate gradient iteration number, the dual problem cost function J, the norm of the gradient of J and the cost $1/2|w|_H^2$ of the primal problem. Effectively Table 3, which describes the conjugate gradient iterations to obtain the sentinel of source 1, shows that it takes many iterations (seven) before convergence (the stop test is gradient norm less than 10^{-2}). Also the sentinel norm |w| is very large. On the contrary, in the case of the second source only one conjugate gradient iteration is enough and the norm of w is not so large.

S = Source
* = Observatory

Figure 2.7

*two sources s1 and s2 and one measurement point *1*

Table 2.3

Source one sentinel: seven conjugate gradient iterations are necessary

Iteration number D	Dual problem cost function	Norm of gradient	Primal problem cost function
1.0000e+00	-7.8513e+06	**5.7449e+02**	7.8513e+06
2.0000e+00	-1.5896e+07	**5.7370e+02**	1.5896e+07
3.0000e+00	-1.6441e+07	**8.5127e+01**	1.6486e+07
4.0000e+00	-1.6441e+07	**1.2216e+01**	1.6486e+07
5.0000e+00	-1.6441e+07	**8.3470e+01**	1.6486e+07
6.0000e+00	-1.6441e+07	**4.6486e+01**	1.6487e+07
7.0000e+00	-1.6441e+07	**2.0139e+02**	1.6487e+07

Table 2.4

Source two sentinel: one conjugate gradient iteration is sufficient

1.0000e+00	-1.2037e+01	**5.7008e-03**	1.2037e+01

The consequence of the fact that the sentinel of the first source is "enorm" is that, since $\lambda_1 = (w^1, z)$, perturbations of the observation z, even very small, are considerably amplified. Already without any noise we do not exactly recover $\lambda_1 = 1$ and $\lambda_2 = 1$, but 1.0556 and 1.0062. With a noise of 20% we recover a reasonable value for λ_2 (7.0160). But λ_1 is completely erroneous (-2.9063e+01!).

To remedy to this situation it is evidently convenient to modify the position of the sensor, or to use several sensors. But, particularly if the sensors are expensive, one prefers to be satisfied with only one sensor, whose emplacement is interactively determined in a quasi-optimal way, thanks to this program to calculate the sentinels and the parameters λ_1 and λ_2. For example, let us place a point of measurement as indicated in Figure 2.8.

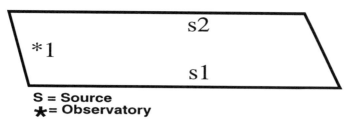

S = Source
★= Observatory

Figure 2.8

Table 2.5

Source one sentinel. calculation of w^1 only needs 3 iterations

cost $J(\gamma)$	Gradient norm	cost $\frac{1}{2}|w|_H^2$
-4.7879e+04	9.9278e-01	4.7879e+04
-9.3804e+04	3.4479e-01	9.5935e+04
-1.0019e+05	1.6044e-02	1.0312e+05

Table 2.6

Source two Sentinel. Calculation of w^2 only needs 3 iterations

Cost	Gradient norm	Cost
-2.4338e+04	5.0464e-01	2.4338e+04
-3.8382e+04	3.4291e-01	3.9021e+04
8.2585e+04	1.7686e-02	5.0804e+04

3 Rectangular domain

3.1 Physical motivations

The aim of this chapter is to provide the reader with a tool enabling him to work with exact solutions for the equations of state. We first derive explicit solutions for the dispersion and reaction equations when the physical domain is a rectangle. We then develop in detail the implementation of the sentinel method in the particular case where Ω is the unit square. We finally present several numerical experiments. Some of them are realized with an observatory that no longer consists of a finite number of points, but is an open set ω contained in Ω: $\omega \subset \Omega$.

3.1.1 First motivation

There are circumstances (a ship discharging a pollutant in the sea, for example) where, when there is only one pointwise source of pollution, only a neighborhood of this point is of concern (at least at the beginning) and one can legitimately suppose that the interesting phenomena take place in a square centered at that point, with a zero concentration condition on the boundary of this domain. This chapter brings some complements in this case.

3.1.2 Second motivation

Another motivation for working on a rectangular domain is that the state equations can be exactly solved and we can explicitly give the value of the pollutant concentration at each point of the domain and at each instant of the time interval. We give these solutions in the case of a linear diffusion-reaction parabolic equation with zero concentration on the boundary (Dirichlet boundary condition). This also enables us to be free of the influence of discretization errors on the results when calculating and evaluating the methods for sentinels. It is interesting to analyze the sentinel method with the assurance that the possible drawbacks do not come from the solution of the state equations.

3.1.3 Third motivation

The physical domain (a laguna for example) effectively may be a rectangle.

3.2 Modeling the physical system

Let Ω be the unit square $\Omega =]\,0,1\,[\times]\,0,1\,[$ and $\Gamma = \partial\Omega$ its boundary. We suppose that there are N_1 sources of pollution, the i-th source being located at point a_i and discharging pollutant at a flow rate that is a function of time of the form $\lambda_i\,s_i(t)$ kilograms per day where the N_1 functions of time $s_i(t)$ are known but the corresponding parameters λ_i are unknown. The function $s_i(t)$ represents the modulation of the i-th source as a function of time, whereas the parameter λ_i is the intensity of the discharge.

The pollutant concentration y is governed by the following conditions

$$\begin{cases} \dfrac{\partial y}{\partial t} - \Delta y + \sigma\, y = \displaystyle\sum_{i=1}^{i=N_1} \lambda_i s_i(t)\delta(x - a_i) & on\ \ Q = \Omega \times]0,T[\\[2mm] y(x,t) = 0 & on\ \Sigma = \Gamma \times]0,T[\\[2mm] y(x,0) = \displaystyle\sum_{j=1}^{N_2} \tau_j w_j(x) & on\ \Omega \end{cases} \qquad (3.2.1)$$

Here, for $j = 1,\ 2,\ \cdots,\ N_2$, w_j is the j-th eigenfunction of the functional operator $-\Delta + \sigma$ operating on functions vanishing on Γ. Let $\left(w^j,\ \mu^j\right)$ be the eigenpairs of this operator. We have

$$\begin{cases} -\Delta w_j + \sigma w_j = \mu_j w_j & on\ \Omega \\ w_j = 0 & on\ \Gamma \end{cases} \quad \left| w_j \right|_{L^2(\Omega)} = 1. \qquad (3.2.2)$$

It is well known that these functions w^j span $L^2(\Omega)$. Due to the linearity of the system (3.2.1), we can write its solution as follows,

$$y(x,t) = \sum_{i=1}^{N_1} \lambda_i \Psi_i(x,t) + \varphi(x,t)$$

where the functions Ψ_i and φ are the solutions of problems

$$\begin{cases} \dfrac{\partial \Psi_i}{\partial t} - \Delta \Psi_i + \sigma \Psi_i = s_i(t)\delta(x - a_i) & on \ Q = \Omega \times \,]0, T[\\ \Psi_i(x, t) = 0 & on \ \Sigma = \Gamma \times \,]0, T[\ . \\ \Psi_i(x, 0) = 0 & on \ \Omega \end{cases}$$

(3.2.3a)

$$\begin{cases} \dfrac{\partial \varphi}{\partial t} - \Delta \varphi + \sigma \, \varphi = 0 & on \ Q = \Omega \times \,]0, T[\\ \varphi(x, t) = 0 & on \ \Sigma = \Gamma \times \,]o, T[\\ \varphi(x, 0) = \displaystyle\sum_{j=1}^{N_2} \tau_j w_j(x) & on \ \Omega \end{cases}$$

(3.2.3b)

The exact solution of (3.2.3) is explicitly given

$$\varphi(x, t) = \sum_{i=1}^{N_2} \tau_j w_j(x) \exp\left(-\mu_j t\right).$$

(3.2.4)

Remark 3.2.1

We indicate in the next section how to obtain the exact solution of (3.2.2). We begin with the one- dimensional case, where $\Omega = \,]0,1[$, then pass to the two-dimensional case, where $\Omega = \,]0,1[\times \,]0,1[$. The method easily extends to the case of a cube $\Omega = \,]0,1[^n$, $n = 3$. An interesting feature from the numerical point of view is that we can by this method access to the exact solution of three-dimensional problems with a Macintosh!

3.3 Exact solution of the direct problem

So we are faced with the problem (3.2.2), solving a dispersion-reaction equation on a segment or a square with a Dirac mass at the right-hand side, zero Dirichlet boundary values, and zero initial values. The reader not interested in mathematical details can directly skip to formula (3.3.18) for the one-dimensional case and to formula (3.3.24) for the two-dimensional case.

3.3.1 One-dimensional case

We are going to explicitly solve, with $0 \le b \le 1$ and $s \in L^2(0,T)$

$$\begin{cases} \dfrac{\partial y}{\partial t} - \dfrac{\partial^2 y}{\partial x^2} + \sigma y = \delta(x-b)s(t) \\ y(0,t) = 0 \\ y(1,t) = 0 \\ y(x,t) = 0 \end{cases} \qquad (3.3.1)$$

Let y(x, t, s, b) be the solution of (3.3.1) for the function of modulation s and the (source) position b.

N.B. The case $\displaystyle\sum_{i=1}^{i=N_1} s_i(t)\big(x-a_i\big)$ is the same: the output resulting from the

input $\displaystyle\sum_{i=1}^{i=N} \lambda_i s_i(t)\delta\big(x-a_i\big)$ is $\displaystyle\sum_{i=1}^{i=N} \lambda_i \Psi_i(x,t)$ .where $\Psi_i(x,t) = y(x, t ; s_i, a_i)$.

Step 1: (facultative)

The change of unknown function

$$y(x,t;b) = [\exp(-\sigma t)]z(x,t;b) \qquad (3.3.2)$$

suppresses the term σy. It appears $s(t)[\exp(\sigma t)]\delta(x-b)$ in the second member.

Step 2: definition of function z

All that remains is to analytically calculate the solution z of

$$\begin{cases} \dfrac{\partial z}{\partial t} - \dfrac{\partial^2 z}{\partial x^2} = s(t)[\exp(\sigma t)]\delta(x-b) \\ z(0,t) = 0 \\ z(1,t) = 0 \\ z(x,0) = 0 \end{cases} \qquad , \; x \in \,]0,1[, \; t > 0, \; 0 < b < 1 \qquad (3.3.3)$$

Step 3: Definition of points b_n

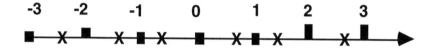

The symbol X indicates the position of numbers - b- 2, b-2, - b, b, - b+2, b+2, etc..., on the real line.

From point b on the axis of real numbers we generate other points either by successive symmetries whose centers are the integer numbers and/or translations of amplitude two. These numbers are the following, where $k \in Z$ (the relative numbers) :

$$b_{2k} = b+2k, \ b_{2k-1} = b_{-1}+2k, \text{ with } b_0 = b \text{ and } b_{-1} = -b \qquad (3.3.4)$$

An algorithm to generate the $4 \times (n+1)$ first points is the following:

b being between 0 and 1, let α and β be the vectors $\alpha = [b \ \ -b \ \ b-2 \ \ 2-b]$ and $\beta = [2 \ \ -2 \ \ -2 \ \ 2]$

$\gamma = \alpha$

for i = 1 to n, $\gamma = \gamma + \beta$, $\alpha = [\alpha \ \gamma]$, end.

Example: b = 0.2 and n = 3 and 4n = 12

$\alpha = [0.2000 \ \ -0.2000 \ \ -1.8000 \ \ 1.8000]$ \qquad\qquad i = 0

$\beta = [2 \ \ -2 \ \ -2 \ \ 2]$

$\gamma = \alpha = [0.2000 \ \ -0.2000 \ \ -1.8000 \ \ 1.8000]$

$\gamma = \gamma + \beta = [2.2000 \ \ -2.2000 \ \ -3.8000 \ \ 3.800]$

$\alpha = [0.2000 \ \ -0.2000 \ \ -1.8000 \ \ 1.8000 \ \ 2.2000 \ \ -2.2000 \ \ -3.800 \ \ 3.8000]$
i = 1

$\gamma = \gamma + \beta = $ [4.2000 -4.2000 -5.8000   5.800]

$\alpha = [\alpha\ \gamma] = $ [0.2000 -0.2000 -1.8000 1.8000 2.2000 -2.2000 -3.8000
3.8000   4.2000 -4.2000 -5.8000   5.800] $i = 2$

$\gamma = \gamma + \beta = $ [6.2000   -6.2000   -7.8000   7.8000]

$\alpha = [\alpha\ \gamma] = $ [0.2000 -0.2000 -1.8000   1.8000   2.2000 -2.2000 -3.8000
3.8000   4.2000 -4.2000 -5.8000   5.8000   6.2000 -6.200 -7.8000
7.800] $i = 3$

$\gamma = \alpha = $ [0.2000   -0.2000   -1.8000   1.8000

$\gamma = \gamma + \beta = $ [2.2000   -2.2000   -3.8000   3.800]

$\alpha = $ [0.2000 -0.2000 -1.8000   1.8000   2.2000 -2.2000 -3.800 3.8000]
$i = 1$

$\gamma = \gamma + \beta = $ [4.2000   -4.2000   -5.8000   5.800]

$a = [a\ g] = $ [0.2000 -0.2000 -1.8000 1.8000 2.2000 -2.2000 -3.8000
       3.80   4.2000 -4.2000 -5.8000   5.800 $i = 2$

$g = g + b = $ [6.2000   -6.2000   -7.8000   7.8000]

$a = [a\ g] = $ [0.2000 -0.2000 -1.8000 1.8000 2.2000 -2.2000 -3.8000
3.8000   4.2000 -4.2000 -5.8000   5.8000   6.2000 -6.200 -7.8000
7.800] $i = 3$

One thus arrives to the definition of a function u on R, whose restriction to]0, 1[is z :

Let us denote $f(t) = s(t)\,e^{\sigma t}$ and a $= b_n$ (3.3.7)

We are now faced with the problem

$$\begin{cases} \dfrac{\partial v}{\partial t} - \dfrac{\partial^2 v}{\partial x^2} = f(t)\delta(x-a) \\ v(x,0) = 0 \end{cases}$$ (3.3.8)

We solve this problem by taking the Fourier transform of both sides with respect to x. Let V denote the Fourier transform of v

$$V(\lambda, t) = \int_{-\infty}^{+\infty} exp(-2i\pi\lambda x)v(x)dx. \qquad (3.3.9)$$

We obtain

$$\begin{cases} \dfrac{\partial V}{\partial t} + 4\pi^2\lambda^2 V = f(t)exp(-2i\pi\lambda a) \\ V(x,0) = 0 \end{cases} \qquad (3.3.10)$$

The solution of this Cauchy problem is given explicitly by

$$V(\lambda, t) = Y(t)\int_0^t f(t-r)exp\left(-4\pi^2\lambda^2 r\right)exp(-2i\pi\lambda a)dr \qquad (3.3.11)$$

Applying the inverse Fourier transform in order to come back to the variable x, we obtain

$$v(x,t) = Y(t)\int_0^t f(t-r)\dfrac{exp\left(-\dfrac{x^2}{4r}\right) *_x \delta(x-a)}{2\sqrt{\pi r}}dr \qquad (3.3.12)$$

where $exp\left(\dfrac{-x^2}{4r}\right) *_x \delta(x-a) = exp\left(\dfrac{-(x-a)^2}{4r}\right)$, ($*_x$ is the convolution product with respect to x).

$$v(x,t) = Y(t)\int_0^t s(t-r)e^{\sigma(t-r)}\dfrac{exp\left(\dfrac{-|x-a|^2}{4r}\right)}{2\sqrt{\pi r}}dr \qquad (3.3.13)$$

whence

$$u(x,t) = Y(t)\sum_{n=-\infty}^{n=+\infty}(-1)^n\int_0^T s(t-\tau)exp(\sigma(t-\tau))\dfrac{exp\dfrac{-(x-b_n)^2}{4\tau}}{2\sqrt{\pi\tau}}d\tau \qquad (3.3.14)$$

Evidently it is enough to keep b_0, b_{-1}, b_1, and eventually some others.

We have successively defined functions y, z, u, v and V, and found an explicit expression to calculate y(x, t, s, b)

$$y(x,t;s,b) = Y(t) \sum_{n=-\infty}^{+\infty}(-1)^n \int_0^t s(t-r)e^{-\sigma r} \frac{exp\left(\frac{-(x-b_n)^2}{4r}\right)}{2\sqrt{\pi r}} dr \qquad (3.3.17)$$

$\forall$ k $\in$ Z, b_{2k} = b+2k and b_{2k-1} = b_{-1}- 2k, with b_0 = b and b_{-1} = b. We do the same in the two-dimensional case.

3.3.2 Two-Dimensional case

We suppose a zero initial condition and a zero boundary concentration. We want to explicitly solve (3.3.18)

Step 1: Introducing the function z

As in the one-dimensional case the change of unknown function y = $e^{-\sigma t}z$ suppresses the term in σ and the term $exp(\sigma t)\delta(x-b)$ s(t) appears on the right hand side. Then all that remains is to analytically solve the system

$$\begin{cases} \dfrac{\partial z}{\partial t} - \Delta z = \exp(\sigma t)s(t)\delta(x-b) \,, x \in \ ]0,1[\times]0,1[, \ t > 0 \\ z(x,t) = 0 \ on \ \Sigma = \Gamma \times]0,T[\\ z(x,0) = 0 \ on \ \Omega \end{cases} \qquad (3.3.19)$$

where $b = (b(1), b(2))$, $0 \le b(i) \le 1$, $i = 1, 2$

Step 2: Introducing Dirac masses at points b_n

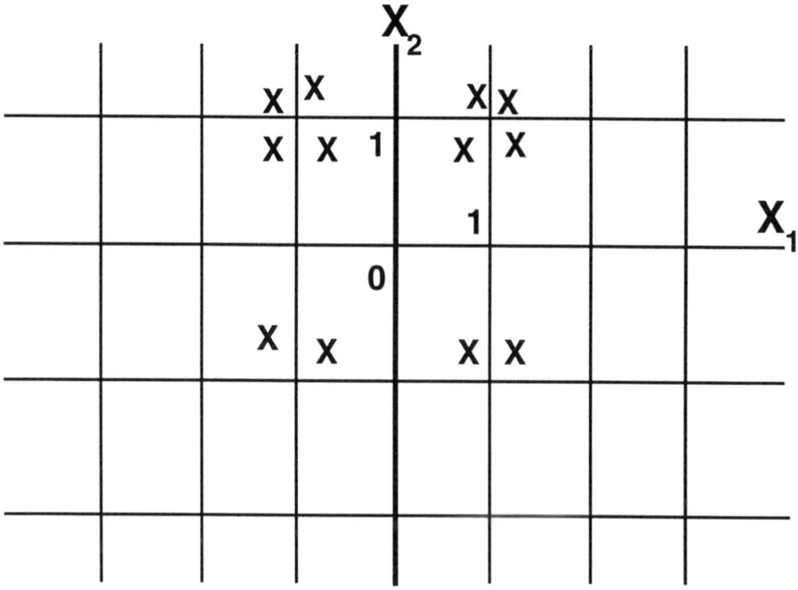

Figure 3.2

By successive symmetries: starting from point b in the square $]0, 1[\times]0, 1[$ introduce the points $b_{n'}$ labelled by a cross X ($b_0 = b$).

Step 3: Introducing function u

Let (Px_1, Px_2) be the coordinates of the point of the square $]0, 1[\times]0, 1[$, which is the original of the point (x_1, x_2). Let us define u on R^2 by $u(x_1, x_2) = z(x_1, x_2)$ on the square $]0, 1[\times]0, 1[$ and alternatively

$$u(x_1, x_2) = \begin{cases} z(Px_1, Px_2) & \text{if } (x_1, x_2) \in \text{ a square labelled } + \\ -z(Px_1, Px_2) & \text{if } (x_1, x_2) \in \text{ a square labelled } - \end{cases}$$

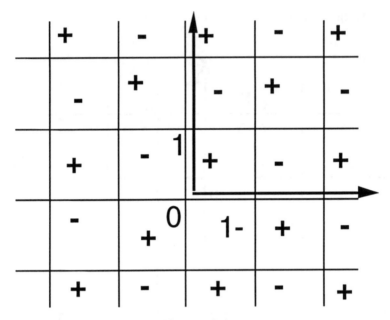

Figure 3.3

Starting from the square $S_0 =]0, 1[\times]0, 1[$, introduce the squares S_n

$$\frac{\partial u}{\partial t} - \Delta u = s(t)(\exp(\sigma t))\sum_{n=-\infty}^{n=+\infty}(-1)^n \delta(x - b_n) \ , \ x \in R^2, \ t \geq 0 \qquad (3.3.20)$$

$$u(x, 0) = 0 \qquad (3.3.21)$$

which implies, by using the Fourier transform in R^2, the following:

$$y(x,t) = Y(t) \sum_{n=-\infty}^{+\infty}(-1)^{n1+n2} \int s(t-r)\exp(\sigma(t-r))\frac{\exp\left(\frac{-|x-b_n|^2}{4r}\right)}{4\pi r} dr$$

$$(3.3.22)$$

where

$x \in R^2$, $b \in R^2$ and $|.|$ is the Euclidian norm in R^2: $|x - b_n|^2 = |x_1 - b_{n_1}|^2 + |x_2 - b_{n_2}|^2$.

the b_{n_1} are defined by $\forall k \in Z$, $b_{2k} = b(1) + 2k$ and $b_{2k-1} = -b(1)-2k$.

the b_{n_2} are defined by $\forall k \in Z$, $b_{2k} = b(2) + 2k$ and $b_{2k-1} = -b(2)-2k$.

$b(1)$ and $b(2)$ are the coordinates of point b and $n = (n_1, n_2)$.

We keep only a few Dirac masses, the series in (3.3.22) converging very quickly.

Table 3.1

values $y(x, t)$ calculated in this way for a Dirac mass placed at point $(0.7, 0.8)$ with $\sigma = 1$, $t = 0.5$ and using 64 Dirac masses.

0	-0.0000	-0.0000	-0.0000	-0.0000	-0.0000
0	0.0038	0.0079	0.0104	0.0079	-0.0000
0	0.0077	0.0160	0.0213	0.0163	-0.0000
0	0.0114	0.0243	0.0328	0.0254	-0.0000
0	0.0148	0.0320	0.0440	0.0342	-0.0000
0	0.0172	0.0378	0.0527	0.0413	-0.0000
0	0.0179	0.0400	0.0564	0.0446	-0.0000
0	0.0165	0.0372	0.0530	0.0421	-0.0000
0	0.0127	0.0289	0.0414	0.0330	-0.0000
0	0.0069	0.0158	0.0228	0.0182	-0.0000
0	-0.0000	-0.0000	-0.0000	-0.0000	0.0000

We thus have an explicit expression (3.3.22) for the function Ψ_i, where b_n is the n-th point generated by successive symmetries from $b = a_i$, as indicated in (3.3.22).

3.4 Sentinels for inversion

Let the system be governed by

$$
\begin{cases}
\dfrac{\partial y}{\partial t} - \Delta y + \sigma\, y = \sum_{i=1}^{i=N_1} \lambda_i s_i(t)\delta(x - a_i) & on \ Q = \Omega \times]0, T[\\[2mm]
y(x,t) = 0 & on \ \Sigma = \Gamma \times]0, T[\\[2mm]
y(x,0) = \sum_{j=1}^{\infty} \tau_j w_j(x) & on \ \Omega
\end{cases}
\qquad (3.4.1)
$$

We approximate the solution y of (3.4.1) by truncating the series defining the initial condition:

$$y = \sum_{i=1}^{N_1} \lambda_i \psi i + \sum_{j=1}^{N_2} \tau j \Phi j \qquad (3.4.2)$$

where ψ_i and φ_j are defined, respectively, by (3.3.24) and

$$\begin{cases} \dfrac{\partial \Phi_j}{\partial t} - \Delta \Phi_j + \sigma \Phi_j = 0 \ on \ Q \\ \Phi_j(x,t) = 0 \qquad\qquad on \ \Sigma \ . \\ \Phi_j(x,0) = w_j(x) \qquad on \ \Omega \end{cases} \qquad (3.4.3)$$

Let us define the exact observation z of the state y as the value of the pollutant concentration at point x_k and at time t_l:

$$z = y(x_k, t_l), \ 1 \le k \le M, \ 1 \le l \le N_t$$

$$y(x_k, t_l) = \sum_{i=1}^{N_1} \lambda_i \Psi_i(x_k, t_l) + \sum_{j=1}^{N_2} \tau_j \Phi_j(x_k, t_l)$$

and let C be the observation operator : z = Cy. To estimate the n-th component λ_n of the vector λ associated with z_d, we need the sentinel associated with this parameter, in order to apply the formula

$\lambda_n = (w^n, z_d)_H$, z_d being the observation. We present the direct method in the case when the concentrations of pollutant are measured at discrete instants.

3.5 Direct method

1. Calculate the numbers $\Psi_i(x_k, t_l)$ and $\Psi_{j+N_1}(x_k, t_l) = \Phi_j(x_k, t_l)$ for $1 \le i \le N$, $1 \le j \le N_2$, $1 \le k \le M$), $.1 \le l \le N_t$, which are the $N \times M \times N_t$ elements of the observation $C\Psi$. We "augment" the vector λ by calling $\lambda_{j+N_1} = \tau_j$ for $1 \le j \le N_2$. We still call λ this augmented vector: $\lambda \in R^N$, $N = N_1 + N_2$

2. Calculate the square $N \times N$ matrix Λ:

$$\Lambda_i^j = \sum_{k=1}^{k=M} \sum_{l=1}^{l=N_i} \Psi_i(x_k, t_l) \Psi_j(x_k, t_l), \quad 1 \le i, j \le N. \tag{3.5.1}$$

3. Solve the linear system $\Lambda \gamma^n = e^n$ where e^n is the n-th vector of the canonical basis of R^N and obtain γ^n, element of R^N, $w^n = B\gamma^n = C\Psi\gamma^n$.

4. We then have the sentinel

$$\tilde{w}^n = B\tilde{\gamma}^n = C\tilde{\rho}^n = C\Psi\tilde{\gamma}^n.$$

Remark 3.5.1

The formula (3.5.1) is to be compared with

$$\sum_{k=1}^{k=M} \int_0^T \Psi_i(x_k, t) \Psi_j(x_k, t) dt, \quad 1 \le i, j \le N$$

3.6 Numerical experiments

For these numerical experiments we take $\sigma = 1$, as functions w_i the modes of $A = -\Delta + \sigma$, for coefficients τ_i the Fourier coefficients of a function f, defined below. The general expression of these eigenfunctions is

$$\begin{cases} w_k = 2^{1/2} \sin(k\pi x_1) & \text{in dimension 1} \\ w_{kl}(x_1, x_2) = w_k(x_1) w_l(x_2) & \text{in dimension 2} \end{cases} \tag{3.6.1}$$

The corresponding eigenvalues are

$$\lambda_{kl} = \sigma + (k^2 + l^2)\pi^2 \tag{3.6.2}$$

The coefficients τ_i are the inner products in $L^2(\Omega)$ of modes w_i and a random pollution distribution f on the domain Ω, generated in the following way. We define a function f depending of four parameters f_1, f_2, f_3 and f_4:

$$f(x_1, x_2) = f_1 x_1 x_2 + f_2(1 - x_1)x_2 + f_3(1 - x_1)(1 - x_2) + f_4 x_1(1 - x_2)$$

$$\tag{3.6.3}$$

where the coefficients f_1, f_2, f_3 and f_4 are the values of f respectively at the four corners of the unit square, (1,1), (0,1), (0,0), and (1,0). These coefficients are generated according to a uniform probability law on [0,a], a > 0. We easily verify that the Fourier coefficients of this initial distribution f are

$$\tau_i = (f, w_i)_{L^2(\Omega)} = \int_\Omega f(x_1, x_2) w_{k,l}(x_1, x_2) dx_1 dx_2$$

$$= \frac{1}{kl\pi^2} \left\{ f_1(-1)^{k+1} + f_2(-1^{l-1}) + f_3 + f_4(-1)^{k-1} \right\}$$

(3.6.4)

For our numerical simulations the initial "missing" data are the τ_i

$$y = y_0(x) = \sum_{j=1}^{j=N_2} \tau_j w_j(x)$$

(3.6.5)

In the following numerical experiments the little rectangles represent the observatory (compare their actual size to the size of the the unit square Ω) and the circles o indicate the position of pointwise sources. The times of observation are t = {0.1, 0.2, 0.3, 0.4, 0.5}(unit of time: one day)

3.6.1 Numerical experiment 1

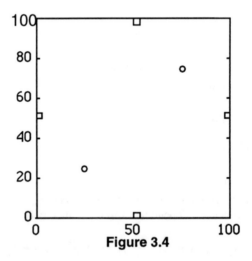

Figure 3.4

Exact value	No noise	Noise 0.01	Noise 0.05	Noise 0.9
1.0000	1.0000	1.0061	1.0353	0.6894
2.0000	2.000	1.9946	1.963	2.3079

Here the sentinel method gives poor results, excepted when the noise is practically absent.

3.6.2 Numerical experiment 2

Two sources of pollution (circles) and an observatory consisting of four rectangles

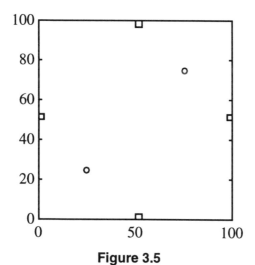

Figure 3.5

Exact value	No noise	Noise 0.001	Noise 0.01	Noise 0.1	Noise 0.4	Noise 0.9
1.0000	1.0000	1.0002	1.0131	0.9792	0.8351	1.1851
2.0000	2.0000	1.9999	1.9866	1.9939	2.1454	1.8567

3.6.3 Numerical experiment 3

The sources are the same as above, but the observations are at points (0.01, 0.01), (0.01, 0.99), (0.99, 0.99) and (0.99, 0.01). It is, paradoxically, a better result than with a non-pointwise observatory! In fact the paradox is only apparent. It is due to the fact that the noise of observation is random.

Exact value	No noise	Noise 0.4
1.0000	1.0000	0.9657
2.0000	2.0000	2.0901

3.6.4 Numerical experiment 4

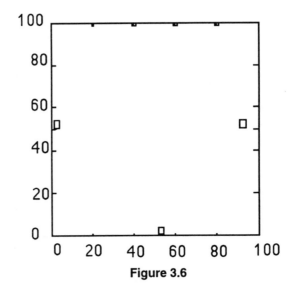

Figure 3.6

Exact value	No noise	Noise 0.001
1.0000e+00	1.0000e+00	1.0029e+00
2.0000e+00	2.0000e+00	1.9920e+00
5.0000e-01	5.0000e-01	5.1010e-01
2.0000e+00	2.0000e+00	1.9943e+00

Here the sentinel method gives poor results, excepted when the noise is practically absent.

3.6.5 Numerical experiment 5

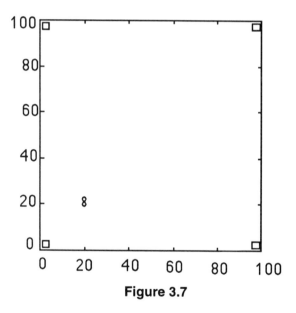

Figure 3.7

With two sources very close to each other, let us choose now as observatory four small squares with edges 0.03, therefore of measure 0.03 × 0.03 = 0.0009 each. It is remarkable that with so small an observatory's surface and with the sources so close to each other one can still discriminate each source's contribution.

Exact value	No noise	Noise 0.01	Noise 0.05	Noise 0.9
1.0000	1.0000	1.0061	1.0353	0.6894
2.0000	2.0000	1.9946	1.9631	2.3079

Exact value	No noise	Noise 0.001	Noise 0.01	Noise 0.05
1.0000	1.0000	0.997	1.01	0.8992
2.0000	2.0000	2.002	1.98	2.0887

3.6.6 Numerical experiment 6

With the sources of experiment 4 and observatories of experiment 3 we find

Exact value	No noise	Noise 0.02
1.0000e+00	1.0000e+00	1.0111e+00
2.0000e+00	2.0000e+00	2.0029e+00

3.6.7 Numerical experiment 7

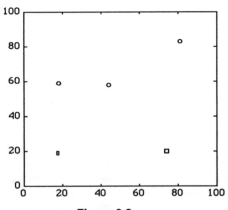

Figure 3.8

Exact value	No noise	Noise 0.01
100.0000	100.0000	101.1447
200.0000	200.0000	196.7807
150.0000	150.0000	152.5445

3.6.8 Numerical experiment 8

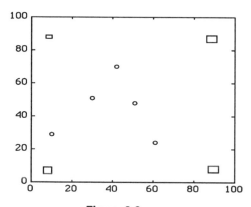

Figure 3.9

Exact value	No noise	Noise 0.05	Noise 0.20
1.0000	1.0000	1.0410	0.8742
2.0000	2.0000	1.9165	2.0001
3.0000	3.0000	3.0831	2.8439
4.0000	4.0000	4.0375	3.8827
5.0000	5.0000	4.9503	5.3651

3.6.9 Numerical experiment 9

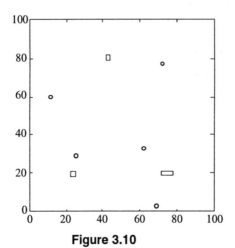

Figure 3.10

Exact value	No noise	Noise 0.05	Noise 0.20
1	1	1.0410	0.8742
2	2	1.9165	2.0001
3	3	3.0831	2.8439
4	4	4.0375	3.8827
5	5	4.9503	5.3651

3.6.10 Numerical experiment 10

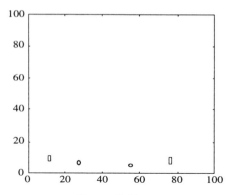

Figure 3.11

Exact value	No noise	Noise 0.05	Noise 0.05	Noise 0.2
1.0000	1.0000	1.1905	0.9390	0.8582
10.0000	10.0000	9.9480	10.0626	10.3113

3.7 Sensitivity to the size of the observatory

As a matter of fact, by continuity all the values observed in ω will be close to those observed at any point of ω. The numerical experiments show that when ω is shrinking to an observatory consisting of a finite number of points, the numerical results stay good. It is not necessary for the observatory surface to be very large. In fact, it may be zero, ω then consisting of a few measurement points judiciously distributed with respect to the position of the sources and their probable intensity. That is good because most of the time we only have pointwise measurements at our disposal, and they are very expensive. Of course, still supposing the observation points distributed "at best", the more they are, the better it is! Furthermore, one feels that it is better to have measurement points well distributed on Ω rather than only one open set ω if this open set is restricted to lie in a small region of Ω.

4 Sentinels in a river

The aim of this Chapter is to show how to apply the sentinel method to a river. In Section 4.1 we present the physical background and model the pollution in a river in the case where this pollution is due to a consumption of oxygen by carbonated wastes, thus reducing the oxygen concentration at the disposal of animal or vegetal species. In Section 4.2 we give an exact expression for the pollutant concentration in the particular case of a rectangular domain. In Section 4.3 we show how to implement the sentinel method in this particular case.

4.1 Oxygen kinetics and polluted water

4.1.1 BOD 5 (Biological Oxygen Demand during 5 days).

Among all the evolution problems, the simplest and more widespread one is the initial value problem governed by the ordinary differential equation $y' + ky = 0$, and the initial value $y(0) = y_0$ where k is a positive parameter, of dimension the inverse of a time, and y represents a concentration, of a pollutant species for example. Of course the solution of this problem is $y(t) = y_0 \exp(-kt)$.

Example 4.1.1: biological oxygen demand

This demand undergoes an exponential decrease as a function of time by the auto-purificationphenomenon. At time $t = t^* = k^{-1}$ this demand is only $y_0/e = y_0/1.732...$ In our river we have k = 0.19, whence $t^* = 5$ days. The BOD y_0 at instant t = 0 of a polluted water sample is usually determined in the laboratory by means of observations during 5 days, whence the name of BOD 5. It is equal to the quantity of dissolved oxygen consumed by a polluted water incubated during 5 days at 20°C. A decrease of dissolved oxygen can rapidly lead to death for many aerobic organisms. The result may be an ill-smelling river contaminated by pathogen.microbes The BOD 5 of surface waters intended to produce water for alimentation should never be more than 3 mg/l.

Example 4.1.2: Nitrates

y is the nitrate concentration and k = 0.14. The consumption term ky must be multiplied by 3.56 to get the equivalent consumption of dissolved oxygen.

4.1.2 BOD and Dissolved Oxygen: The model of Streeter and Phelps

Let y_1 denote the BOD and y_2 the dissolved. oxygen concentration. The following system of differential equations is called the Streeter and Phelps model[1]:

$$
\begin{cases}
y_1' + k_1 y_1 = 0 \\
y_2' + k_2(y_2 - y_2^*) = k_1 y_1 \\
y_1(0) = y_0^1 \\
y_2(0) = y_0^2
\end{cases}
\tag{4.1.1}
$$

y_2^* is the temperature-dependent equilibrium value, in milligrams per liter, of dissolved oxygen in the absence of pollution. We evidently have, $D = y_2^* - y_2$ denoting the demand in oxygen and $L = y_1$ the load in biological oxygen, the system of two ordinary differential equations

$$
\begin{cases}
L' + k_1 L = 0, \ L(0) = L_0 \\
D' + k_2 D = k_1 L, \ D(0) = D_0
\end{cases}
\tag{4.1.2}
$$

whose solution is

$$
\begin{cases}
L(t) = L_0 \exp(-k_1 t) \\
D(t) = L_0 \dfrac{k_1}{k_2 - k_1}(\exp(-k_1 t) - \exp(-k_2 t)) + D_0 \exp(-k_2 t)
\end{cases}
\tag{4.1.3}
$$

4.1.3 BOD, nitrates, and dissolved oxygen: the model of krenkel

This is a model taking into account the BOD load L, dissolved oxygen D, and nitrates N:

$$
\begin{cases}
L' + k_1 L = 0 & L(0) = L_0 \\
D' + k_2 D = k_1 L + 3.56 k_3 N & D(0) = D_0 \\
N' + k_3 N = 0 & N(0) = N_0
\end{cases}
\tag{4.1.4}
$$

We easily obtain a solution of the form

$$D(t) = c_1 \exp(-k_1 t) + c_2 \exp(-k_2 t) + c_3 \exp(-k_3 t) \tag{4.1.6}$$

4.1.4 Parameter values

We are interested in the pollution discharge in a section of a river. We adopt as length and time units the meter and the day. This section is a rectangle of dimensions L_1 = 4620 m and L_2 = 60 m. The flow velocity v is supposed to be constant and equal to 7 km /d. The longitudinal dispersion coefficient E_x is 8 $m^2 s^{-1}$. We suppose that the transversal dispersion coefficient E_z is given by the Okubo Law[1]

$$E_z = E_x \left(\frac{L_2}{L_1} \right)^{\frac{4}{3}} \tag{4.1.7}$$

where L_1 and L_2 are the dimensions of the rectangle in the longitudinal and transverse directions. Then we have:

Table 4.1: known parameters

L1 = 4.62 km = 4620 m	longitudinal dimension
L2 = 60 m	transverse dimension
L3 = 1.7 m	depth of the section
D1 = Ex × 24 × 3600 = 691200m2 d-1	longitudinal dispersion coefficient = 700, 000
D2 = 2.110e+03 m2d-1	transversal dispersion coefficient = 2000
km/d = 7000 m d-1	velocity
k = 0.19 d-1	reaction coefficient

Table 4.2: characteristic times

convection	Time_Conv=L1 / v = 16 h
longitudinal dispersion	Time_Diff_Long = L12/D1 =1 month
transversal dispersion	Time_Diff_Transverse = L22/D2 41 h
reaction	Time_React = 1 / k = 5.26 d

4.2 Convection-dispersion-reaction equation

Take L_1 and L_2 as new length units in, respectively, the longitudinal and transverse directions. In an evolution regime the pollutant concentration y is governed by the equation

$$\frac{\partial y}{\partial t} + c\frac{\partial y}{\partial x_1} - D_L\frac{\partial^2 y}{\partial x_1^2} - D_T\frac{\partial^2 y}{\partial x_2^2} + ky = \frac{1}{L_1 L_2 L_3}\sum_{i=1}^{i=N_1}\lambda_i s_i(t)\delta(x - a_i),$$

$$0 \le x_i \le 1$$

(4.2.1)

where the coefficients are

$$D_L = \frac{D_1}{L_1^2} = 3.2383\text{e-}02, \quad D_T = \frac{D_2}{L_2^2} = 5.8611\text{e-}01, \quad c = \frac{v}{L_1} = 1.5152\text{e+}00, \quad k = 1.9000\text{e-}01$$

where the BOD y is expressed as a function of time in kilograms of oxygen per cubic meter (or, which is the same, in milligrams of oxygen per liter), t in days, c in L_1 per day, x1 in L_1, x2 in L_2, D_L in D_L in L_1^2 per day, and D_T in L_2^2 per day.

Let x denote the point of coordinates (x_1, x_2), y = y(x, t) = y(x₁, x₂, t). Ω = the strap R ×] 0, 1[, and the boundary of Ω: Γ = R ×{0, 1}, Q = Ω×]0, T[, Σ = Γ ×]0,T[.

4.2.1 The right hand side

$$\frac{1}{L_1 L_2 L_3}\sum_{i=1}^{i=N_1}\lambda_i s_i(t)\delta(x - a_i)$$

is a term of production (in kilograms per cubic meter and per day). The coefficient λ_i is the intensity of the i-th source of pollution (in kilograms of oxygen per day) .The dimensionless values $s_i(t)$ of the function s_i are known and indicate how the flow rate of these sources of pollution varies with time. The i-th source of pollution is located at point a_i and $\delta(x - a_i)$ is the Dirac mass at this point.

4.2.2 Boundary conditions

We assume that the riverside is impervious to the pollutant

$$\frac{\partial y}{\partial n} = 0 \text{ for } x \in \Gamma \text{ and } t \in]0, \text{ T}[\text{ where } \frac{\partial y}{\partial n} = \begin{cases} \dfrac{\partial y}{\partial x_2} & \text{if } x_2 = 1 \\ -\dfrac{\partial y}{\partial x_2} & \text{if } x_2 = 0 \end{cases}. \tag{4.2.2}$$

We solve the delicate problem of the upstream and downstream boundary conditions by placing these limits respectively at $+\infty$ and $-\infty$. That is to say that the space variable x_1 lies between $-\infty$ and $+\infty$. Let us recall that for the considered section x_1 a priori lies between $x_1 = 0$ and $x_1 = 1$

4.2.3 Initial conditions

We approach the initial condition by a function that is constant on each element of a rectangular mesh, each rectangle $[a_j, b_j] \times [c_j, d_j]$ bearing a pollution of density τ_j ($j = 1, ..., N_2$). We suppose that these rectangles are placed between $x_1 = 1$ and $x_1 = +1$.

$$y(x, 0) = \sum_{j=1}^{N_2} \tau_j \chi_j \tag{4.2.3}$$

where χ_j is the characteristic function of the j-th rectangle: $\chi_j = \chi_{(a_j, b_j)} \times \chi_{(c_j, d_j)}$.

The missing data are the values τ_j (supposed to be constant by rectangle) of the initial pollutant concentration on the rectangle (a_j, b_j, c_j, d_j), $1 \le j \le N_2$ (essentially above the section we are interested in and on this section).

4.3 Exact solution of state equation

We begin by taking z instead of y as the unknown function, where y and z are related by $y = \exp(\alpha x_1 + \beta t) z$. It is easy to determine the coefficients α and β so that the coefficients of z and $\dfrac{\partial z}{\partial x_1}$ are zero. We find

$$\begin{cases} \alpha = \dfrac{c}{2D_L} \\ \beta = -k - \dfrac{c^2}{4D_L} \end{cases} \tag{4.3.1}$$

and we now have to solve for z the system

$$
\begin{cases}
\dfrac{\partial z}{\partial t} - D_L \dfrac{\partial^2 z}{\partial x_1^2} - D_T \dfrac{\partial^2 z}{\partial x_2^2} = \dfrac{1}{L_1 L_2 L_3} \sum_{i=1}^{i=N_1} \lambda_i s_i(t)\delta(x - a_i)\exp(-\alpha x_1 - \beta t) \\[3mm]
\dfrac{\partial z}{\partial n} = 0 \\[3mm]
z(x,0) = \displaystyle\sum_j^{j=N_2} \tau_j \exp(-\alpha x_1)\chi_j(x)
\end{cases}
\tag{4.3.2}
$$

Due to the fact that the solution z of (4.3.2) depends linearly upon the N_1 parameters λ_i and the N_2 parameters τ_j we can express this solution as

$$
z = \sum_{i=1}^{i=N_1} \lambda_i \psi_i + \sum_{j=1}^{j=N_2} \tau_j \varphi_j
$$

where ψ_i is the solution of

$$
\begin{cases}
\dfrac{\partial \psi_i}{\partial t} - D_L \dfrac{\partial^2 \psi_i}{\partial x_1^2} - D_T \dfrac{\partial^2 \psi_i}{\partial x_2^2} = \dfrac{1}{L_1 L_2 L_3} s_i(t)\delta(x - a_i)\exp(-\alpha x_1 - \beta t) \\[3mm]
\dfrac{\partial \psi_i}{\partial n} = 0 \\[3mm]
\psi_i(x,0) = 0
\end{cases}
\tag{4.3.3}
$$

and Φ_j is the solution of the problem

$$
\begin{cases}
\dfrac{\partial \varphi_j}{\partial t} - D_L \dfrac{\partial^2 \varphi_j}{\partial x_1^2} - D_T \dfrac{\partial^2 \varphi_j}{\partial x_2^2} = 0 \\[3mm]
\dfrac{\partial \varphi_j}{\partial n} = 0 \\[3mm]
\varphi_j(x,0) = \exp(-\alpha x_1)\chi_j(x)
\end{cases}
\tag{4.3.4}
$$

Let us first solve the problem (4.3.3)

4.3.1 Pointwise source of pollution

$$\begin{cases} \dfrac{\partial z}{\partial t} - D_L \dfrac{\partial^2 z}{\partial x_1^2} - D_T \dfrac{\partial^2 z}{\partial x_2^2} = \dfrac{1}{L_1 L_2 L_3} s_i(t)\delta(x-b)\exp(-\alpha x_1 - \beta t) \\ \dfrac{\partial z}{\partial n} = 0 \\ z(x,0) = 0 \end{cases}$$ (4.3.5)

1. Introduce, by successive symmetries starting at point $b \in \Omega$, the points b_n of coordinates $(b_{n,1}, b_{n,2})$ (represented by a circle o in Figure 4.1)

$b^0 = b,\ b^{n,1} = b^{0,;1} \ \forall\, n$ and $b^{2k,2} = b^{0,2} + 2k,\ b^{2k-1,2} = - b^{0,2} + 2k,\ k = 0, + 1, -1, + 2, -2,$ etc...

2. Define the straps

$\Omega^0 = \Omega,\ \ \Omega^n = R \times\,]n,n+1[,\ n = 0, +1, -1, +2, -2,$ etc.

3. The domain of definition of the function z is restricted to Ω. Define its extension u to R^2:

$u(x, t) = z(x, t)$ on Ω^0, $u(x, t) = z\,(2\text{-}x, t)$ on Ω^1, $u(x, t) = z(\text{-}x, t)$ on Ω^{-1}, etc.

(continue by successive symmetries and/or translations parallel to the x_2 axis and of amplitude two).

We thus arrive at the problem: find u on R^2 such that

$$\begin{cases} \dfrac{\partial u}{\partial t} - D_L \dfrac{\partial^2 u}{\partial x_1^2} - D_T \dfrac{\partial^2 u}{\partial x_2^2} = \dfrac{1}{L_1 L_2 L_3} s(t)\sum_{n=-\infty}^{n=+\infty}\delta\!\left(x-b^n\right)\exp(-\alpha b_1 - \beta t) \\ u(x,0) = 0 \end{cases}$$ (4.3.6)

By using the Fourier transform as in Chapter 3, we find

$$\Psi_i(x,t) = Y(t)\frac{1}{4\pi L_1 L_2 L_3 \sqrt{D_L D_T}}$$

$$\sum_{n=-\infty}^{n=+\infty}\int_0^t s_i(\tau)\exp\big(-k(t-\tau)\big)\frac{1}{t-\tau}\exp\left(\frac{-\big(x_1-b_1-c(t-\tau)\big)^2}{4D_L(t-\tau)}\right)\exp\left(\frac{-\big(x_2-b_{n,2}\big)^2}{4D_T(t-\tau)}\right)d\tau$$

(4.3.7)

and it is enough to keep a few Dirac masses.

4.3.2 Response to an initial condition

When the initial condition is the characteristic function χ_j of a rectangle

$$[a_j,b_j]\times[c_j,d_j],\quad -\infty \le a_j \le b_j \le +\infty,\ \ 0\le c_j \le d_j \le 1$$

$$\Phi_j(x,t)=\frac{1}{4}\exp(-kt)erf(A_1,B_1)\sum_{n=-\infty}^{n=+\infty}erf(A_2,B_2)+erf(C_2,D_2)$$

(4.3.8)

where A_1 and B_1 are the values of

$$X=\frac{\xi_1-x_1+ct}{2\sqrt{D_L t}}$$

(4.3.9)

respectively for $\xi_1 = a_j$ and for $\xi_1 = b_j$, and where A_2, B_2, C_2, and D_2 are the values of

$$X=\frac{\xi_1-x_2}{2\sqrt{D_T t}}$$

(4.3.10)

respectively for $\xi_2 = c_j + 2n$, $\xi_2 = d_j + 2n$, $\xi_2 = -c_j + 2n$, and $\xi_2 = -d_j + 2n$ and where $erf(X_1, X_2)$ is the value of the error function integrated from X_1 to X_2:

$$erf(X_1,X_2)=\frac{2}{\sqrt{\pi}}\int_{X_1}^{X_2}\exp\big(-t^2\big)dt$$

(4.3.11)

Remark 4.3.1

There is no problem because this series converges. For example, in the extreme case where the initial pollutant distribution is uniformly equal to 1 on the section, and therefore by successive symmetries all over R^2, one easily finds y(x, t) = exp(- kt), which is the result one naturally expect.on [aj, bj] × [cj, dj].

Remark 4.3.2

We do not need many rectangles to obtain a sentinel insensitive to initial conditions. Practically we took four rectangles.

Remark 4.3.3

If the initial pollution rectangles are rectangles [aj, bj] × [0, 1] bearing a pollution density μ_j, a tedious but straightforward calculation gives the j-th rectangle contribution

$$y_j(x,t) = \frac{1}{2}\mu_j \exp(-kt)\,\mathrm{erf}(X_1, X_2)$$

where X_1 and X_2 are the values of

$$X = \frac{(\xi_1 - x_1 + ct)}{2\sqrt[2]{D_L\,t}}$$

respectively for $\xi_1 = a_j$ and for $\xi_1 = b_j$.

4.4 Sentinels for a river (evolution regime)

From the above sections it results that the pollutant concentration y in a river can be described by

$$\begin{cases} y' + Ay = \sum_i \lambda_i s_i(t)\delta(x - a_i) \\ y(0) = \sum_j \tau_j \chi_j(x) \end{cases}$$

$$\chi^j(x) = \begin{cases} 1 \text{ if } x \in [aj, bj] \times [cj, dj] \\ 0 \text{ otherwise} \end{cases}$$

We measure y at points x_k, $(1 \le k \le M)$ and at each instant $t \in \,]0, T[$.

4.4.1. Direct method to obtain a sentinel

To obtain a sentinel by the direct method, we have to solve a linear system of matrix Λ.

1. Let Ψ include the N_1 "elementary solutions" Ψi and the $N2$ "elementary solutions" Φ_j and let $N=N1+N2$. For each source number i, $1 \leq i \leq N$, calculate the observation $C\Psi_i$ due to this source. This provides the operator $B = \left(C\psi_1 \ C\psi_2 \ \cdots \ C\psi_N \right)$

2. Calculate the components of matrix Λ. These components are provided by the inner products in $L^2(0, T;R^M)$ of the N functions $\Psi_i(x_k, t)$, $1 \leq k \leq M$ and $t \in]0, T[$:

$$\Lambda_i^j = \left(C\Psi_i, C\Psi_j \right)_H = \Lambda_i^j = \sum_{k=1}^{k=M} \int_0^T \Psi_i(x_k, t)\Psi_j(x_k, t)dt$$

$$1 \leq i, j \leq N1 \leq i, j \leq N$$

3. Calculate γ^n solution of the linear equation

$$\Lambda\gamma^n = e^n$$

4. Then the n-th sentinel w^n is given by

$$w^n = B\gamma^n$$

4.5 Numerical experiments

4.5.1 Example 1

Let us consider, to fix ideas, a river section like the one that is represented by Figure 4.1.

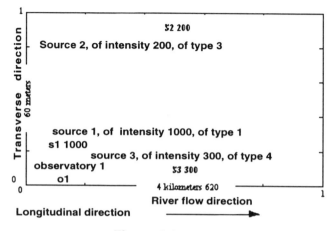

Figure 4.1

Water flows from the left to the right at a velocity of 7 km/h. The length of the section is 4km620, its width 60 m. We have taken these lengths as units in, respectively, the longitudinal and transversal directions, which explains the coordinates from 0 to 1 in these two directions. The notation s2 200 means that here the pollution source number 2 intensity is 200. This source is exactly located at the point under the s of s2 200. Similarly o1 and o2 are the points of measurement, point-wise too (that corresponds well to the actuality of measurements).

A priori we do not know the intensity of pollutant flow. In fact, each polluter has a tendency to place on others the responsibility for a pollution. In order to identify the intensity of a source we have at our disposal information resulting from measurements. Figure 4.2 shows the two functions of time obtained by measuring the pollutant concentration at each of the two observatories o1 and o2.

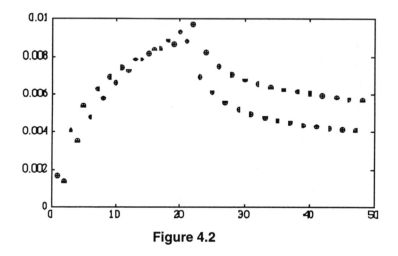

Figure 4.2

Pollutant concentration observed at each of two points of measurement as a function of time. In the abscissa we have the time of the observation. There are two observations per hour during 24 h. therefore 48 observations, in the order: at time 1, y(1,1) and y(1,2), at time 2, y(2,1) and y(2,2), ..., at time 24, y(24,1,) and y(24,2). These observations are concentrations expressed in kilograms per cubic meter, which corresponds to grams per liter, or 1000 mg/l. The pollutant at this concentration, 0.01, therefore corresponds to 10 mg/l.. Since in order to be potable, surface waters are required to have pollution less than 3 mg/l, we see that the pollution of our system is serious.

Let us suppose that we dispose of exact measures, those indicated in Figure 4.2 for example. Then the method of sentinels gives estimated values, 1000, 200, and 300, identical to the exact values. But in fact the hypothesis of a complete absence of measurement noise is unreallistic. We therefore introduce a slight noise. When measurements are polluted by noise, the exact values of the parameters and their estimation by the method of sentinels cannot be the same. However, the fit of the estimated to the exact values is still fairly satisfying. With a noise of 5% we find

1007.2   211.2   284.7

instead of

1000.00   200.000   300.000

Before dealing with inverse problems, let us examine results obtained by a numerical analysis of state equations. We first treat direct problems. Let us consider a point, observatory number 1 for example. We obtain the results represented by Figure 4.3.

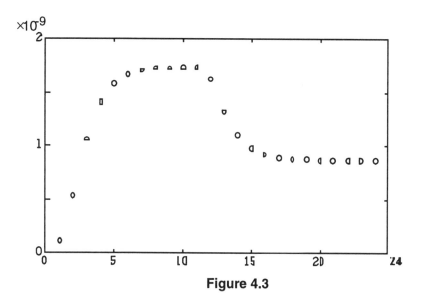

Figure 4.3

Pollution at (0.04, 0.07), due to source 1

Source 1 has greater influence than sources 2 or 3 upon this observatory (concentration on the order of 9 to 10 g/l in response to 1kg/j: in consequence, the response to the 1000 kg of source 1 on a day will be on the order of 0.001 mg/l). We can neglect the responses to the sources 2 and 3. That is natural when one considers the position of observatory 1 with respect to sources 1, 2 and 3 (figure 4.2).

Presented here is a representation by contours, which are isovalues of the concentration, and arrows, which represent the sum of the dispersion flux and the convection flux. We represent, the response to a single source of 1 kg/d.

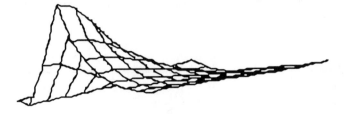

Figure 4.5

Pollution after1 h by source 1.

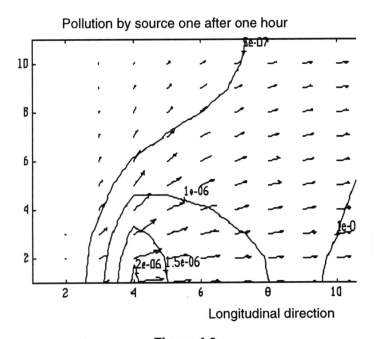

Figure 4.6

Pollution by source 1 after 24 h.

4.5.2 Example 2

Let us consider *a real case* on a river section with many food industries, cities with important pollutions, and the efflux of several small rivers.

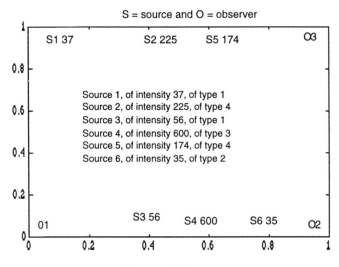

Figure 4.7

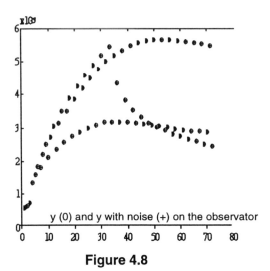

Figure 4.8

Table 4.3: Comparison of exact values (first line) and estimated values (second line)

Noise = 0

225.0000	56.0000	600.0000	174.0000	35.0000 exact values
225.0000	56.0000	600.0000	174.0000	35.0000 estimated values

Noise = .05

225.0000	56.0000	600.0000	174.0000	35.0000
238.2965	65.2007	673.5155	168.3192	-2.4572

Noise = 0.2

225.0000	56.0000	600.0000	174.0000	35.0000
216.5887	53.1747	581.7944	177.1227	44.3705

5 Our first nonlinear problem

5.1 Linear case

5.1.1 Physical motivation

Let us consider a reservoir of water with the same quantity of ingoing and outgoing water. Suppose an unknown initial concentration of pollutant in the reservoir and an unknown concentration of pollutant in the ingoing water. Our aim is to identify these two unknown parameters, by observating the pollutant concentration in the outgoing water during some time interval. This reservoir could be a lake or a compartment in a biochemical model.

5.1.2 Modelling of the physical system

In the case of an (admittedly over simplified) model of a lake, we assume that the change of pollution level in the lake is due to the ingoing pollutant minus the outgoing pollutant. Let y be the pollution level (i.e., the concentration of pollutant) in the lake with the same incoming and outgoing flow rates, so that the volume of the lake is constant. The input (and output) pollutant concentrations are, respectively, s_0 and y. The inverse τ of parameter ρ is a known time characteristic of the lake: $\tau = \rho^{-1}$..

It is the time it takes, when $s_0 = 0$, to pass from $y(0) = y_0$ to $y(\rho^{-1}) = e^{\frac{y_0}{}}$. The system is governed by the ordinary differential equation

$$\frac{dy}{dt} + \rho(y - s_0) = 0 \qquad (5.1.1)$$

on the time interval $]0,T[$, with the initial condition

$$y(0) = y_0. \text{ at } t = 0 \qquad (5.1.2)$$

5.1.3 Statement of the identification problem

The parameter $\tau = \rho^{-1}$ is known, but the two parameters y_0 and s_0 are unknown. We wish to identify them, not necessarily y_0, but at least s_0.

Indeed, we are not interested in knowing the initial concentration y^0 and will be satisfied if we can identify the input concentration s_0.

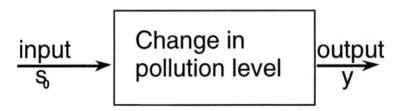

Figure 5.1

The state of the system is the solution y of (5.1.1) and (5.1.2) on the time interval $]\,0, T[$. It depends linearly upon the two parameters y_0 and s_0, so we can explicitly write the solution as a linear combination of ψ_1 and ψ_2,

$$y(t) = y_0\psi_1(t) + s_0\psi_2(t) \tag{5.1.3}$$

ψ_1 and ψ_2 solutions of (5.1.1) and (5.1.2), respectively, for ($s_0 = 0$ and $y_0 = 1$) and ($s_0 = 1$ and $y_0 = 0$).

$$\begin{cases} \psi_1(t) = \exp(-\rho t) \\ \psi_2(t) = 1 - \exp(-\rho t) \end{cases} \tag{5.1.4}$$

The problem is to identify s_0 by using the data at our disposal. These data consist of the observation of y on a time interval $]\,t_0, t_1[\,\subset\,]\,0, T[$. It is important to note here that the quality of the observation depends upon the choice of this interval.

$H = L^2(t_0, t_1)$ is the space of observations, equipped with the inner product $(f, g)_H = \displaystyle\int_{t_0}^{t_1} f(t)g(t)dt$.

$|\cdot|_H$ is the norm associated with this inner product:

$$|f|_H^2 = (f, f)_H = \int_{t_0}^{t_1} f^2(t)dt$$

$$V = R^2$$

is the space of parameters, equipped with the usual Euclidean inner product:

$$(u,v)_V = \sum_{i=1}^{i=2} u_i v_i$$

We indifferently write

$$v = \begin{bmatrix} v_1 \\ v_2 \end{bmatrix} = \begin{bmatrix} y_0 \\ s_0 \end{bmatrix} \text{ or } v = (v1, v2) = (y0, s0)$$

Let B: V → H be the operator from V into H that associates with the vector of parameters v the observation

$$z(t) = y_0 \psi_1(t) + s_0 \psi_2(t) \quad , t \in \big] t_0, t_1 \big[\tag{5.1.5}$$

It is important to distinguish between the state y, that is a function defined on the time interval $]0,T[$, and its observation z, that is a function defined on the time interval $]t_0, t_1[$. The latter is the restriction of the former to the time interval $]t_0, t_1[$.

Let Cy denote this observation of y. We can rewrite Cy(t) = z(t) as the product B(t)v where B(t) is the (time dependent) 1× 2 matrix $\begin{bmatrix} \psi_1(t) & \psi_2(t) \end{bmatrix}$ and v is the column vector $\begin{bmatrix} v_1 \\ v_2 \end{bmatrix} = \begin{bmatrix} y_0 \\ s_0 \end{bmatrix}$. Alternatively,

$$y = \begin{bmatrix} \psi_1 & \psi_2 \end{bmatrix} \begin{bmatrix} v_1 \\ v_2 \end{bmatrix} \text{ if we view y, } \psi_1, \text{ and } \psi_2 \text{ as functions of time on}$$
$$]0,T[\text{ and}$$

$$z = \begin{bmatrix} C\psi_1 & C\psi_2 \end{bmatrix} \begin{bmatrix} v_1 \\ v_2 \end{bmatrix} \text{ if we view Cy,C}\psi_1, \text{ and C}\psi_2 \text{ as functions of}$$
$$\text{time on }]t_0, t_1[$$

It results that z(t) = B(t) v where B(t) = $\begin{bmatrix} C\psi_1(t) & C\psi_2(t) \end{bmatrix}$ or z = Bv where $\begin{bmatrix} C\psi_1 & C\psi_2 \end{bmatrix}$.

Unfortunately, the observation of y is never exact. Because of noise in the measurements, errors due to the observer and crude approximations in the modelling, the observation is $z_d(t)$ instead of $Cy(t)$:

$$z_d(t) = Cy(t) + error(t) \text{ for } t \in \left] t_0, t_1 \right[.$$ (5.1.6)

As a consequence there is no reason for z_d to be of the form of z in (5.1.5).

5.1.4 Least squares

The usual way to overcome this difficulty is to identify the parameters v in the least squares sense, i.e., to minimize a cost function J which is the square of the distance between the calculated solution $y(v)$ of (5.1.1) and (5.1.2) and the actual observation. A noisy measurement z_d of this state $y(v)$:

$$J(v) = \frac{1}{2} |Cy(v) - z_d|_H^2, v \in V$$ (5.1.7)

where

$$v = \begin{bmatrix} y_0 & s_0 \end{bmatrix}^T$$

is the vector of unknown parameters and $y(v)$ is the solution of (5.1.1) and (5.1.2).

It is well known that u minimizes (5.1.7) then

$$\Lambda u = B^* z_d$$ (5.1.8)

where $\Lambda = B^* B$ is the symmetric 2×2 matrix whose elements are

$$\Lambda_i^j = \left(Be^i, Be^j \right)_H 1 \le i, j \le 2.$$ (5.1.9)

It is a simple exercise to prove that the operator B is one-to-one from V into H. $Bv = 0 \Rightarrow v = 0$, so that the matrix Λ is strictly positive definite and in particular Λ^{-1} exists. Then (5.1.8) is equivalent to

$$u = \Lambda^{-1} B^* z_d \tag{5.1.10}$$

By (5.1.10) to each $z_d \in H$ there corresponds an element u *of* V. We can write (5.1.10) as

$$u = W z_d \tag{5.1.11}$$

where the operator W is the so-called pseudo inverse of B:

$$W = \Lambda^{-1} B^* \tag{5.1.12}$$

We easily check that

$$WB = \mathrm{Id}_V \tag{5.1.13}$$

(identity operator in V), whence also the name of left inverse of B given to W.

If e^i denotes the i-th vector of the canonical basis of V, the i-th component of u is given by

$$u_i = (e^i, u)_V = (e^i, W z_d)_V = (W^* e^i, z_d)_H = (B \Lambda^{-1} e^i, z_d)_H = (w^i, z_d)_H$$

Definition   5.1.1

w^i is called the sentinel associated with the i-th parameter.

Proposition   5.1.1

Given $z_d \in H$, the least squares estimation of the i-th parameter, u_i , is the

inner product in H of the sentinel w^i and the observation z_d:

$$u_i = (w^i, z_d)_H \quad ., \quad w^i = B \gamma^i \ , \quad A \gamma^i = e^i . \tag{5.1.14}$$

$$u = [\begin{smallmatrix} u_1 \\ u_2 \end{smallmatrix} \ ] = \begin{bmatrix} (w^1, z_d)_H \\ (w^2, z_d)_H \end{bmatrix} \tag{5.1.15}$$

Comparing (5.1.11) and (5.1.15), we identify the pseudo inverse operator W and the two sentinels: $\begin{bmatrix} w^1 \\ w^2 \end{bmatrix}$

$$W = \begin{bmatrix} w^1 \\ w^2 \end{bmatrix} \tag{5.1.16}$$

Remark 5.1.1

Suppose that we observe the absence of pollutant in the reservoir during the time interval $]t_0,t_1[$, that is, $y(t) = 0$ for t in $]t_0,t_1[$. This implies that there is no supply of pollutant, neither an initial concentration nor an input later. Hence $y_0 = 0$ and $s_0 = 0$.

Remark 5.1.2

An alternate way to say that B is one-to-one is $Bu = Bv \Rightarrow u = v$, which means that a given observation corresponds to only one vector of parameters. This uniqueness property is fundamental in inverse problems.

5.1.5 Direct method to calculate sentinels

From the above discussion the following algorithm appears to be a natural candidate to determine a sentinel. We call it the "direct method":

1. Compute B:

$$Be^i = C\Psi_i, \ 1 \le i \le 2 \tag{5.1.17}$$

2. Calculate the components of matrix Λ:

$$\Lambda_{i,j} = \int_{t_0}^{t_1} C\Psi_i(t) \, C\Psi_j(t) \, dt \tag{5.1.18}$$

3. Calculate γ^i:

$$\Lambda\gamma^i = e^i \tag{5.1.19}$$

4. Define w^i by

$$w^i = B\gamma^i \tag{5.1.20}$$

5.1.6 Estimating a parameter

For a given observation z_d, the i-th parameter u_i is given by the inner product in H

$$u_i = (w^i, z_d)_H \tag{5.1.21}$$

In particular s_0 is given by

$$s_0 = u_2 = (w^2, z_d)_H = \int_{t_0}^{t_1} w^2(t) z_d(t) dt \tag{5.1.22}$$

Remark 5.1.3

The identification of parameter u_i is accomplished in two steps. In the first step we calculate the sentinel w^i by (5.1.17) and (5.1.20). The sentinel w^i is determined once for all. The observation z_d plays no role in this step. The second step consists in calculating u_i by the inner product (5.1.21). This step is very fast.

An advantage of the sentinels method is this possibility to pre calculate a sentinel, then use it repeatedly for a fast treatment of incoming new data. Another advantage of the sentinels method is that it indicates how much each parameter value is perturbated by noise: $\left|w^i\right|_H$ gives this information.

5.1.7 A first program: direct method

We suppose that y(t) is known on the time interval (t_0, t_1), together with the parameter ρ. On the other hand, we ignore the values of y_0 and s_0. Because of the linearity of the operator B we explicitly have the solution y of (5.1.1) and (5.1.2) as a linear combination of ψ_1 and ψ_2, solutions of (5.1.1) and (5.1.2), respectively, when ($s_0 = 0$ and $y_0 = 1$) and ($s_0 = 1$ and $y_0 = 0$): $y = y_0\psi_1 + ys_0\psi_2$. This solution is represented in Figure 5.2.

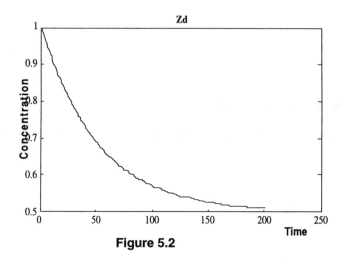

Figure 5.2

Program 5.1.1

Here is a program to identify s_0 and y_0.

Param denotes the vector of parameters. param = (s_0, y_0 and ρ)

```
Clear
param = [1 2 3]                          % exact values of the
                                           parameters
dt = 01 ; T = 2 ; t = [0:dt:T]           % initializations
phi = exp (-param (3)*t) ; psi = [phi; 1-phi]   % calculation of psi
plot (psi'), title ('psi'), xlabel ('time'),
  ylabel ('concentration')
Pause
y = param (1:2)*psi                      % state of the system
n1 = 90 ; n2 = 120
obs_time_interval = [n1 n2]
zd = y (n1:n2)                           % observation zd
plot (zd), title ('zd'), xlabel ('time'), ylabel
  ('concentration')
pause Lambda = psi (:, (n1:n2))*psi(:,   % building Lambda
  (n1:n2))'*dt ;
disp ('condition number of Lambda'), cond
  (Lambda)
X = Lambda\eye (size (Lambda)) ; pause   % solving Lambda* X =
                                           Identity ;
sentinels = X'*psi (:, (n1:n2)) ; pause  % defining sentinels
```

plot (sentinels'), title ('sentinels'), xlabel
  ('time'), ylabel ('sentinel')
Pause
sentinels_norm = [] ;   % norm of sentinels
for   i   =   1:2,sentinels_norm   =
  [sentinels_norm norm (sentinels (i,:))] ;
  end
par = sentinels*zd'*dt ;   % the parameters
% exact parameter values, reconstructed
  values and norms of sentinels
[param (1:2)' par sentinels_norm']

5.1.8 Numerical experiments

Experiments 5.1

This program gives, in the absence of noise, the exact values of the
parameters. The sensibility of the estimated parameters to noise in the
measurements depends upon the norm of sentinels. Tables 5.1 and 5.2 give
these norms for the sentinels of y_0 and s_0.

Table 5.1

identified parameter values and norms of the sentinels: We can observe
that the error on y_0 should be 20 times larger than the error on s_0.

	Exact values	Identified values	Sentinel norm
y_0	1	1	1406
s_0	2	2	70

Table 5.2

*Norms $\left|w^1\right|$ and $\left|w^2\right|$ of the sentinels of y_0 and s_0: for an observation on the
time interval ($n_1 dt$, $n_2 dt$) The norm of a sentinel indicates how much this
estimated parameter is sensible to noise. We see that the worse situation is
when the interval of observation is (1.9, 2).*

Experiment number	n^1	n^2	$\left\|w^1\right\|$	$\left\|w^2\right\|$
1	10	200	21	15
2	190	200	5679	954
3	90	120	366	202

We thus have at our disposal a tool that tells us those parameters we can expect to be identifiable and those for which the situation is hopeless.

The following figures are relative to an observation on time interval (90, 120).

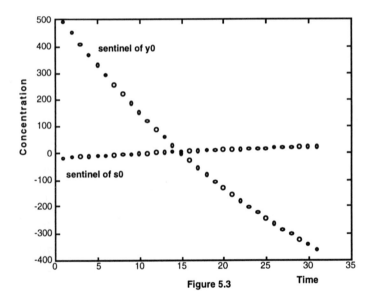

Figure 5.3

Restriction of the state to the time interval (0.90, 1.20), of length 0.30.

5.2 Nonlinear case

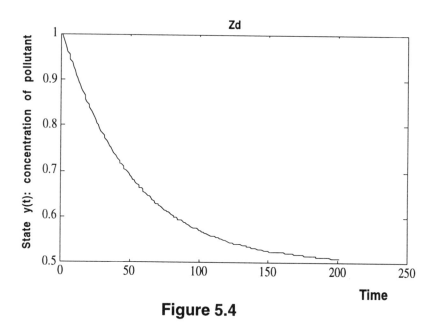

Figure 5.4

We still suppose that we are given the observation of $y(t)$ in some time interval $]t_0, t_1[\subset] 0, T[$ .But we do not know the three parameters y_0, s_0, and ρ . The problem is no more linear in the sense that the observation does not depend linearly upon the vector v of parameters. Now the vector of unknown parameters is

$$v = \begin{bmatrix} y_0 \\ s_0 \\ \rho \end{bmatrix}. \quad v_1 = y_0 \, , \ v_2 = s_0 \ \text{and} \ v_3 = \rho \qquad (5.2.1)$$

Let V, H, and B, respectively, denote the space of parameters the space of observations z, and the (nonlinear) operator that associates this observation z with the parameters v.

$$V = R^3 \ \text{and} \ H = L^2(t_0, t_1), \ B: V \to H$$

$$z(t;v) = v_1 \exp(-v_3 t) + v_2 \left[1 - \exp(-v_3 t)\right]$$

(5.2.2)

Our aim is to identify v from noisy measurements z_d. To achieve this goal we are going to minimise the distance in H of z_d to the image of V by the non linear operator B. We will use for that a descent method involving the sentinels of the linearized problem. So at first we define what we mean by linearized problem, then determine the sentinels of this linear problem.

5.2.1 Linearized system

Let y be the solution of problem (5.1.1) - (5.1.2) for the three parameters

$$v = \begin{bmatrix} y_0 \\ s_0 \\ \rho \end{bmatrix}$$

and $y + \hat{y}$ the solution of the same problem for the three parameters

$$v + \hat{v} = \begin{bmatrix} y_0 + \hat{y}_0 \\ s_0 + \hat{s}_0 \\ \rho + \hat{\rho} \end{bmatrix}$$

where

$$\hat{v} = \begin{bmatrix} \hat{y}_0 \\ \hat{s}_0 \\ \hat{\rho} \end{bmatrix}$$

is a "small" variation of v. We have

$$(y + \hat{y})' + (\rho + \hat{\rho})(y + \hat{y} - s_0 - \hat{s}_0) = 0$$

(5.2.3)

$$y' + \rho(y - s_0) = 0$$

(5.2.4)

Substacting (5.2.4) from (5.2.3) we obtain

$$\hat{y}' + \rho(\hat{y} - \hat{s}_0) + \hat{\rho}(y - s_0) + \hat{\rho}(\hat{y} - \hat{s}_0) = 0$$

Dropping the non linear term $\hat{\rho}(\hat{y} - \hat{s}_0)$ we obtain the so-called "linearized equation"

$$\hat{y}' + \rho(\hat{y} - \hat{s}_0) + \hat{\rho}(y - s_0) = 0. \tag{5.2.5}$$

The initial conditions are, respectively,
$y(0) = y_0$ and $(y + \hat{y})(0) = y_0 + \hat{y}_0$, so that

$$\hat{y}(0) = \hat{y}_0 \tag{5.2.6}$$

Definition 5.2.1

The linearized system of the system (5.1.1)-(5.1.2) is the system (5.2.5)-(5.2.6).

Intuitively, if $\hat{v}$ is "small", the solution $\hat{y}$ of this linearized problem is the principal part of the perturbation to y due to the variation $\hat{v}$ of v. The calculated response to this change is $\hat{z} = C\hat{y}$, that is, $\hat{y}$ restricted to the time interval (t_0, t_1). Since the linearized system (5.2.5)-(5.2.6) is linear with respect to $\hat{v}$, its solution is a linear combination of the solutions obtained when all the components of $\hat{v}$ are 0, excepted one that is equal to 1. We have

For $\hat{y}_0 = 1$, $\hat{s}_0 = 0$ and $\hat{\rho} = 0$, $\psi_1' + \rho\psi_1 = 0$,   and $\Psi_1(0) = 1$

For' $\hat{y}_0 = 0$, $\hat{s}_0 = 1$ and $\hat{\rho} = 0$, $\Psi_2' + \rho(\Psi_2 - 1) = 0$,   and $\Psi_2(0) = 0$

For $\hat{y}_0 = 0$, $\hat{s}_0 = 0$ and $\hat{\rho} = 1$, $\Psi_3' + \rho\Psi_3 + (y - s_0) = 0$,   and $\Psi_3(0) = 0$

Let $B'(v)$[1] be the differential application of B at (v) transforming $\hat{v}$ into $\hat{z}$. Provided it is injective, one can attach to it a generalized inverse, i.e., a family $W(v)$ of three sentinels $w^i(v)$ as we did in Section 5.1. Due

---

[1] Recall that now $V = R^3$

to the linearity of this linearized system, the explicit solution of (5.2.5) and (5.2.6) is

$$\hat{y}(t,v) = \sum_{i=1}^{i=3} \hat{v}_i \psi_i(t,v) \quad \text{where} \quad \begin{cases} \psi_1(t,v) = \exp(-\rho t) \\ \psi_2(t,v) = 1 - \exp(-\rho t) \\ \psi_3(t,v) = -(y_0 - s_0)t\exp(-\rho t) \end{cases}$$

5.2.2 Sentinels of the linearized system

In Section 5.1 we only had $N = 2$ parameters to identify and $N = 2$ associated functions ψ_i. Following the same procedure, we define when $N = 3$ a $N \times N$ matrix $\Lambda(v)$ by the formula

$$\Lambda_i^j = \int_{t_0}^{t_1} C\psi_i(t)C\psi_j(t)dt, \ 1 \le i, j \le 3 \qquad (5.2.7)$$

Then the steps to define the n-th sentinel are

1. Define γ^n by $\Lambda(v)\gamma^n = e^n$

2. Define w^n by $w^n = C\psi\gamma^n = \sum_{i=1}^{i=3}\gamma_i^n \psi_i$ on $]t_0, t_1[$

3. $W(v)$ is defined by the 3 sentinels $w^n(v), 1 \le n \le 3 = N$:

$$W(v) = \begin{bmatrix} w^1(v) \\ w^2(v) \\ w^3(v) \end{bmatrix}$$

It is the generalized inverse of the linear operator $B'(v)$

$$W(v)B'(v) = Id_V \quad {}^2.$$

---

2 $V = R^3$ and Id_V is the Identity operator in V.

The Jacobian operator $B'(v)$ plays the role B was playing in the linear case

Remark 5.2.1

The intuitive nature of the sentinel $w^n(v)$ is the following: Suppose that the components of the vector v are the parameters of the non-linear system and y(v) and z(v) the corresponding state and observation of the state. If one observes a small deviation $\hat{z} \in L^2(t_0, t_1)$ of the observation z (a signal which consists of a function of time defined on (t_0, t_1)), then it can be caused by a small variation $\hat{v}$ of the vector v of parameters, given for the i-th parameter by

$$\hat{v}_i = \left(w^i(v), \hat{z}\right)_V, \ 1 \le i \le 3 \text{ , and } \hat{v} = \left(W(v), \hat{z}\right)_H = \begin{pmatrix} \left(w^1(v), \hat{z}\right)_H \\ \left(w^2(v), \hat{z}\right)_H \\ \left(w^3(v), \hat{z}\right)_H \end{pmatrix}$$

for the 3 parameters.

5.2.3 Least squares

As in the linear case, we look for values of the unknown parameters by using the least squares method. Our problem is to solve, in the least squares sense, the equation

$$B(v) = z_d. \tag{5.2.11}$$

In effect, for z_d given anywhere in H, there is not in general a v in V such that (5.2.11) holds. However, it is always possible to minimize $\left|B(v) - z_d\right|_H$, provided the image of V by B is closed in H. Let the cost function J be defined by

$$J(v) = \frac{1}{2} \left|B(v) - z_d\right|_H^2, v \in V. \tag{5.2.12}$$

We want to find $\inf_{V} J$ where $B(v) = Cy(v)$ is the observation *calculated* by solving (5.1.1) and (5.1.2) and z_d is the *measured* concentration.

We cannot escape the use of differentials and derivatives. So in the following we recall some notions about them.

5.2.4 Differentials and derivatives

1. First, let us consider the so called cost function J defined by $J(z$

$$\frac{1}{2}|z - z_d|_H^2 .$$

We call it a cost function when we view it as a cost for z to be far from z_d..

$$J(z + \hat{z}) - J(z) = (z - z_d, \hat{z})_H + \frac{1}{2}|\hat{z}|_H^2 = (z - z_d, \hat{z})_H + \text{higher order terms}$$

(h.o.t.)

Dropping the h.o.t. we have $J(z + \hat{z}) - J(z) \approx (z - z_d, \hat{z})_H$.

The term $(z - z_d, \hat{z})_H$ is the principal part of $J(z + \hat{z}) - J(z)$ as $\hat{z} \to 0$. It is called the *differential* of J at z and the factor $z - z_d$ in $(z - z_d, \hat{z})_H$ is called the derivative of J at z: we note $dJ = (z - z_d, \hat{z})_H$.

2. Suppose now that z is a non-linear function of v : $z = B(v)$. Then we define $B'(v)$ by

$$B(v + \hat{v}) - B(v) = B'(v)\hat{v} + \text{h.o.t.}$$

Remark 2.2

In the case where $v \to B(v)$ is linear, the left hand side of (5.2.16) is $B\hat{v}$ so that $B'(v)\hat{v} = B\hat{v}$

3. Suppose that $J(v) = \frac{1}{2}|z - z_d|_H^2$ together with $z = B(v)$. The derivatives of respectively J as a function of z and z as a function of v are $J'(z) = z - z_d$ and $B'(v)$. By the *chain-rule* the differential of the function $v \to J(B(v))$ is the appplication

$$\hat{v} \rightarrow \left(J'(B(v)),B'(v)\hat{v}\right)_H = \left(B'^*(v)(B(v)-z_d),\hat{v}\right)_V$$

and we can write this differential as the inner product in V of the gradient J'(v) and $\hat{v}$:

$dJ = \left(B'^*(v)(B(v)-z_d),\hat{v}\right)_V = \left(J'(v),\hat{v}\right)_V$ so that the derivative (or gradient) of J is

$$J'(v) = B'^*(v)(B(v)-z_d) \in V^3$$

4. Fourth, if u minimizes J, it is a solution of the equation J'(u) = 0, which consists of a system of N = 3 (non-linear) equations in N = 3 unknowns u_1, u_2, and u_3.

$$\begin{cases} u \in R^N \\ B'(u)^T(B(u)-z_d) = 0 \end{cases} \tag{5.2.13}$$

We adopt, to minimize J, the following iterative method.

5.2.5 A descent method

$$\begin{cases} v^0 \text{ arbitrary} \\ p^n = -\alpha^n W(v^n)(B(v^n)-z_d) \\ v^{n+1} - v^n = p^n \end{cases} \tag{5.2.14}$$

where the α^n either are chosen optimally or are chosen equal to 1. Applying the formula

$$W = \Lambda^{-1}B^* \text{ with } W = W(v^n), \Lambda = \Lambda(v^n) \text{ and } B = B'(v^n) \text{ we obtain}$$

$$W(v^n) = \Lambda^{-1}(v^n)B'(v^n) \tag{5.2.15}$$

---

[3]When we write $dJ = J'(v)\hat{v}$ it must be understood as $\nabla J(v)\hat{v}$. where

$\nabla J(v) = \left(\dfrac{\partial J}{\partial v_1}\dfrac{\partial J}{\partial v_2}\dfrac{\partial J}{\partial v_3}\right)$ We can also write $dJ = \left(J'(v),\hat{v}\right)_V$.

From (5.2.13), (5.2.14) and (5.15) it results that

$$p^n = \alpha^n \Lambda\left(v^n\right)^{-1}\left(-J'\left(v^n\right)\right) \tag{5.2.16}$$

and

$$-J'\left(v^n\right) = \alpha^n \Lambda\left(v^n\right) p^n \tag{5.2.17}$$

The matrix Λ^{-1} being symmetric positive definite, from (5.2.16) it results that

$$\left(-J'\left(v^n\right), p^n\right) = \left(J'\left(v^n\right), \Lambda^{-1}\left(v^n\right)J'\left(v^n\right)\right) > 0$$

except if $J'\left(v^n\right) = 0$, in which case $v^n = u$, u being a solution of (2.13). Therefore either $J'\left(v^n\right) = 0$ or p^n is a descent direction, that is to say

$$J(v^{n+1}) < J(v^n) \tag{5.2.18}$$

Remark 5.2.3

The fact that the distance between B(v) and z_d is minimum for v = u is characterized by the orthogonality of $B(u) - z_d$ and the vectors $B'(u)\hat{v}, \hat{v} \in V$, which span the plane tangent to the variety $z = B(v)$ at (u, B(u). This orthogonality is equivalent to J'(u)=0:

$$B(u) - z_d \perp B'(u)\hat{v}, \forall \hat{v} \in V \Leftrightarrow (B(u) - z_d, B'(u)\hat{v})_H = 0 \;\; \forall \hat{v} \in V \Leftrightarrow J'(u) = 0 \tag{5.2.19}$$

5.2.6 Nonlinear Case Algorithm

To identify y_0, s_0, and ρ we start with an initial estimation of the parameters $v = v^0$ chosen "arbitrarily" or "at best", given what we know about the parameters. We iterate until convergence. At each iteration, v being given, one achieves the following operations:

1. Calculate the state $y(t,v)$ and the observation $z(t, v) = B(v) = Cy(v)$.

Knowing $(v, y(v))$ calculate the sentinels of the system linearized around this point:

$$W = \{w^i(v), 1 \le i \le N = 3.$$

For that:

2. Calculate $\psi_1(v), \psi_2(v)$ and $\psi_3(v)$, solutions of the linearized problem:

$$\begin{cases} \hat{y}' + \rho(\hat{y} - \hat{s}_0) + \hat{\rho}(y - s_0) = 0 \\ \hat{y}(0) = \hat{y}_0 \end{cases}$$

when the parameters $\hat{y}_0, \hat{s}_0$ and $\hat{\rho}$ respectively are

$\hat{y}_0 = 1,\ \hat{s}_0 = 0$ and $\hat{\rho} = 0,\ (\psi_1)$

$\hat{y}_0 = 0,\ \hat{s}_0 = 1$ and $\hat{\rho} = 0,\ (\psi_2)$

$\hat{y}_0 = 0,\ \hat{s}_0 = 0$ and $\hat{\rho} = 1,\ (\psi_3)$

We thus obtain $\Psi = \begin{vmatrix} \psi_1(v) \\ \psi_2(v) \\ \psi_3(v) \end{vmatrix}$

3. Calculate the 3×3 matrix

$$\Lambda(v) = \int_{t_0}^{t_1} \Psi(v)\Psi^*(v)\,dt.$$

4. Calculate the 3×3 matrix $X(v)$

$$\Lambda(v)X(v) = \mathrm{Id}_V$$

5. Calculate the three sentinel functions

$$w^i(v), 1 \le i \le 3 \text{ by } W(v) = X^*\Psi(v)$$

6. Update the parameter values

$$v^{n+1} = v^n - W(v^n)\left(B(v^n) - z_d\right)$$

5.2.7 Convergence

The fact that we are using a descent method does not imply convergence. However it is a fixed point method, $v = f(v)$ where

$$f(v) = v - W(v)\left(B(v) - z_d\right).$$

It is well-known that, provided it exists a constant L, $0 \le L < 1$ such that $\left|f'(v)\right|_V \le L$, for v in a neightborhood of u, the iterates converge to a vector u such that $u = f(u)$. This is equivalent to $W(u)\left(B(u) - z_d\right) = 0$.

But $W(u) = \Lambda^{-1}(u)B'(u)$ so that $\begin{cases} u \in R^3 \\ J'(u) = B'(u)^*\left(B(u) - z_d\right) = 0 \end{cases}$

$$f'(v)\hat{v} = \hat{v} - \left(W'(v)\hat{v}\right)\left(B(v) - z_d\right) - W(v)B'(v)\hat{v} =$$
$$\hat{v} - \left(W'(v)\hat{v}\right)\left(B(v) - z_d\right) - \hat{v} = -\left(W'(v)\hat{v}\right)\left(B(v) - z_d\right)$$

$$\left|f'(v)\right|_{L(V,V)} \le \left|W'(v)\right|\left|B(v) - z_d\right|_H \quad \text{(Cauchy- Schwarz inequality)}$$

For $\left|B(v) - z_d\right|_H$ small enough, $\left|f'(v)\right|_{L(v,v)} \le L < 1$.

5.2.8 Program

```
clear
disp ('various initializations')
param = [1 1 1] ;% y0, s0, rho in this order
par = [0 0 0] ;% initial estimations
dt = 01 ;
T = 2 ;
```

```
t = [0:dt:T] ;
% generation of zd
disp ('generation of zd')
y = [] ;
temp = param (1) ;
for tt = t,
y = [y temp] ;
temp = (param (2) + temp/dt)/(1/dt + 1 + param (3)) ;
end
zd = y ;
plot  (zd), title ('zd'), xlabel (time), ylabel ('concentration')
pause  (5)
normecart = 1 ;
iteration = 0;
tol = 1.e-7 ;
while normecart > tol,
% generation of z and psi
disp ('generation of z and psi')
y = [];y1 = [] ; y2 = [] ; y3 = [] ;
temp = par (1) ;
temp1 = 1 ;
temp2 = 0 ;
temp3 = 0 ;
for tt =t,
y = [ y1 temp1] ;
y2 = [y2 temp2] ;
y3 = [y3 temp3] ;
temp = (par(2) + temp/dt)/(1/dt + 1 + par(3)) ;
temp1 = (temp1/dt)/(1/dt + 1 + par(3)) ;
```

```
temp2 = (1 + temp2/dt)/(1/dt + 1 + par(3)) ;
temp3 = (-temp1 + temp3/dt)/(1/dt + 1 + par(3)) ;
end
z = y ;
% calculation of psi
disp ('calcul ation of psi')
psi = [y1;y2;y3];
plot (z), title('z'), xlabel(time), ylabel ('concentration')
pause (5)
plot (psi'), title ('psi'), xlabel (time), ylabel ('concentration')
Calculation of A (here denoted A)
A = psi*psi'*dt ;
% solve AX = I for X
disp ('solve AX = I for X')
X = A\eye(A) ;
% calculate sentinels
w = X'*psi ;
plot (w'), title ('sentinels w'), xlabel (time), ylabel ('concentration')
% update parameters
disp ('update parameters')
iteration = iteration+1
ecart = (w*(z-zd)'*dt)'
normecart = norm (ecart)
par = par-ecart
end
```

5.2.9 Identification experiments

Experiment 5.2

Table 5.3

from left to right, at each iteration from 1 to 7, the table indicates the values, from top to bottom, respectively of the iteration number k, the norm $\left|v^{k+1} - v^k\right|_v$, *and the three parameter values stored in the vector* v^k.

Columns1 through 7

1.0000	2.0000	3.0000	4.0000	5.0000	6.0000	7.000	iteration number
1.2399	0.2960	0.1761	0.0977	0.0519	0.0268	0.0136	$\left\|v^{k+1}-v^k\right\|$
0.9864	0.9964	0.9990	0.9997	0.9999	1.0000	1.0000	v_1
0.5893	0.7504	0.8613	0.9265	0.9621	0.9807	0.9903	v_2
0.4658	0.7140	0.8508	0.9236	0.9613	0.9805	0.9902	v_3

Experiment 5.3

Table 5.4

From left to right, at each iteration from 1 to 5, the table indicates the values, from top to bottom, respectively of the iteration number k, the cost $J\left(v^k\right)$, *the norm* $\left|v^{k+1} - v^k\right|_v$, *the norm of sentinel i* $1 \leq i \leq 3$ *at iteration k,* $w^{i,k}$, *and the three parameter values stored in the vector* v^k.

No noise. Observation time interval (0.2, 2)

1.0000	2.0000	3.0000	4.0000	5.0000	Iteration number
2.4327	0.5232	0.1909	0.0149	0.0000	Cost
1.4642	0.2807	0.0194	0.0011	0.0000	Norm (vk+1-vk)
4.9523	12.7283	10.7147	11.6197	11.6915	Norm (sentinel 1)
6.6027	1.2801	1.4405	1.3564	1.3507	Norm (sentinel 2)
35.0485	76.7065	33.7182	36.9856	37.1228	Norm (sentinel 3)
1.2148	1.0031	1.0030	1.0000	1.0000	Parameter 1
1.7485	1.9997	1.9996	2.0000	2.0000	Parameter 2

5.2.10 Observation limited to a time subinterval

Let us now suppose the observation to be effective only in the time interval (0.5, 1.40) (instead of (0,T) where $T = 2$): the changes to the program are minor and consist in defining the time levels from $n1 = 50$ to $n2 = 140$ for z_d, phi and w:

$n1 = 50$; $n2 = 140$; $zd = y(n1:n2)$; $z = y(n1:n2)$;psi $= [y1(n1:n2); y2(n1:n2); y3(n1:n2)]$

Experiment 5.4

Let v be the exact vector of parameters and v_n the approximate vector at iteration n. We have indicated for each iteration n the value of the ratio

$$\frac{\left|v^{k+1} - v^k\right|_H}{\left|v^k - v^{k-1}\right|_H}.$$

We see that the method is of order 1 because

$$\lim_{k\to\infty} \frac{\left|v^{k+1} - v^k\right|_H}{\left|v^k - v^{k-1}\right|_H} = \text{\textbackslash F } (1;2)$$

implies that

$$\lim_{k\to\infty} \frac{\left|v^{k+1} - v\right|_H}{\left|v^k - v\right|_H} = \text{\textbackslash F } (1;2)$$

Table 5.5

1.0000	2.0000	3.0000	4.0000	5.0000	6.0000	7.0000	k
1.1919	0.2891	0.1882	0.1137	0.0644	0.0347	0.0181	$\left\|v^{k+1}{-}v^k\right\|$
1.1919	0.2425	0.6509	0.6043	0.5661	0.5385	0.5211	$\dfrac{\left\|v^{k+1} - v^k\right\|_H}{\left\|v^k - v^{k-1}\right\|_H} \cong \dfrac{1}{2}$
0.9559	0.9822	0.9935	0.9979	0.9994	0.9998	1.0000	Parameter 1
0.5807	0.7257	0.8358	0.9079	0.9507	0.9744	0.9869	Parameter 2
0.4118	0.6605	0.8126	0.9005	0.9485	0.9738	0.9868	Parameter 3
8.0000	9.0000	10.0000	11.0000	12.0000	13.0000	14.0000	k
0.0092	0.0047	0.0023	0.0012	0.0006	0.0003	0.0001	$\left\|v^{k+1}{-}v^k\right\|$

0.5111	0.5057	0.5029	0.5015	0.5007	0.5004	0.5002	$\dfrac{\left|v^{k+1}-v^{k}\right|_{H}}{\left|v^{k}-v^{k-1}\right|_{H}}\cong\dfrac{1}{2}$
1.0000	1.0000	1.0000	1.0000	1.0000	1.0000	1.0000	Parameter 1
0.9934	0.9967	0.9983	0.9992	0.9996	0.9998	0.9999	Parameter 2
0.9934	0.9967	0.9983	0.9992	0.9996	0.9998	0.9999	Parameter 3

Experiment 5.5

Table 5.6

Now $n_1 = 190$ and n_2 $n_2 = 200$. Therefore we observe the state during the final sub-interval (1.9, 2) of the interval (0, 2). We obtain the following results:

Columns 1 through 7

1.0000	2.0000	3.0000	4.0000	5.0000	6.0000	7.0000	k
0.9547	0.1642	0.1614	0.1545	0.1414	0.1213	0.0957	$\left|v^{k+1}-v^{k}\right|$
0.9547	0.1720	0.9831	0.9570	0.9154	0.8581	0.7891	$\dfrac{\left|v^{k+1}-v^{k}\right|_{H}}{\left|v^{k}-v^{k-1}\right|_{H}}$
0.7934	0.8281	0.8643	0.9003	0.9335	0.9608	0.9802	parameter 1
0.5105	0.5800	0.6497	0.7185	0.7842	0.8439	0.8945	parameter 2
0.1465	0.2912	0.4322	0.5658	0.6865	0.7885	0.8674	parameter 3
8.0000	9.0000	10.0000	11.0000	12.0000	13.0000	14.0000	k
0.0686	0.0446	0.0266	0.0149	0.0079	0.0041	0.0021	$\left|v^{k+1}-v^{k}\right|$
0.7166	0.6505	0.5971	0.5583	0.5329	0.5177	0.5092	$\dfrac{\left|v^{k+1}-v^{k}\right|_{H}}{\left|v^{k}-v^{k-1}\right|_{H}}\cong\dfrac{1}{2}.$
0.9916	0.9969	0.9990	0.9997	0.9999	1.0000	1.0000	parameter 1
0.9338	0.9613	0.9786	0.9887	0.9942	0.9970	0.9985	parameter 2
0.9225	0.9573	0.9774	0.9883	0.9941	0.9970	0.9985	parameter 3

Columns 15 through 19

15.0000	16.0000	17.0000	18.0000	19.0000	k
0.0011	0.0005	0.0003	0.0001	0.0001	$\left|v^{k+1}-v^{k}\right|$
0.5047	0.5024	0.5012	0.5006	0.5003	$\dfrac{\left|v^{k+1}-v^{k}\right|_{H}}{\left|v^{k}-v^{k-1}\right|_{H}}\cong\dfrac{1}{2}$

1.0000	1.0000	1.0000	1.0000	1.0000	Parameter 1
0.9993	0.9996	0.9998	0.9999	1.0000	Parameter 2
0.9992	0.9996	0.9998	0.9999	1.0000	Parameter 3

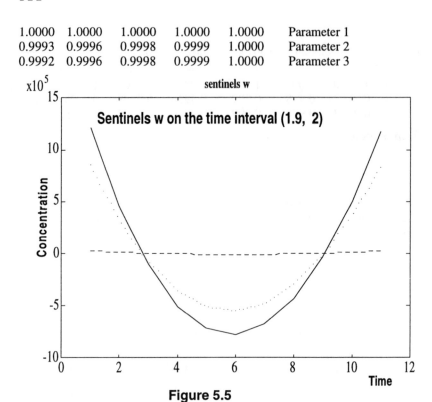

Figure 5.5

The sentinels take their values in interval [-1,000,000, +1,500,000].

The three obtained sentinels (Figure 5.5) are to be compared with those obtained when the time interval during which we observe is (0, T), T = 2 (Figure 5.6). Here the sentinel values are on the order of 10^6. Any noise on the observation will have a catastrophic consequence on the estimation of parameters Consequently our interest is to work with time intervals as long as possible. Anyway it is common sense!

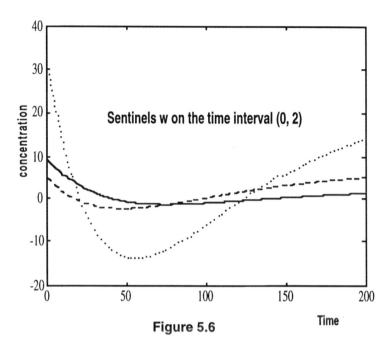

Figure 5.6

The sentinels take their values in interval [-20, +40].

6 Nonlinear problems

This chapter deals with the identification of the time history of pollution release in the case of nonlinear reaction. As is often in numerical analysis, the solution to a nonlinear problem is obtained by successively solving linear problems. Using the sentinels of the linearized problem leads us to introduce a Gauss-Newton algorithm.

6.1 Position of the problem

We return to the problem treated in chapter 1, section 4, where the aim was to identify the evolution of the flow rate as a function of time. We no longer suppose that the reaction term is a linear function of the concentration y, but rather that it obeys a so called Michaelian law of behavior :

$$F(y) = \sigma \backslash F(y; 1+|y|)$$ (6.1.1)

where σ is a positive constant. This is motivated by the fact that indeed most of the time the reaction rates are not linear. The reaction rate given in (6.1.1) is a good representative of such nonlinear reaction rates.

We suppose that the concentration y(x, t) of pollutant, at point $x = (x_1, x_2)$ and time t, is a solution of the nonlinear parabolic equation

$$\frac{\partial y}{\partial t} - \Delta y + F(y) = s(t)\delta(x-a) \quad for\ x\ \in \Omega\ and\ t \in\,]\,0, T[.$$ (6.1.2)

There is only one source of pollution, located at point a, and its flow rate is an unknown function of time. We have a zero flux boundary condition

$$\frac{\partial y}{\partial n} = 0$$. (6.1.3)

The initial condition is

$$y(x, 0) = y^0(x)$$ (6.1.4)

As in chapter 1, section 4, the flow rate s(t) can be approached with a linear combination of hat functions s_i

$$s(t) = \sum_{i=1}^{i=N_1} \lambda_i s_i(t).$$ (6.1.5)

The function s is a piecewise linear function, whose value at instant $t = i\,\Delta t$ is λ_i. Similarly, we approach the initial condition y^0 with a linear combination of characteristic functions χ_j of a finite elements partition of Ω

$$: y^0(x) = \sum_{j=1}^{j=N_2} \tau_j \chi_j(x), \, x \in \Omega. \tag{6.1.6}$$

$$\lambda = (\lambda_1 \, ..., \, \lambda_i, \, ..., \, \lambda_{N_1})^T \in R^{N1} \tag{6.1.7}$$

represents the pollution parameters and

$$\tau = (\tau_1, \, ..., \, \tau_i, \, ..., \, \tau_{N_2})^T \in R^{N2}$$

represents the missing data parameters. Let

$$v = (\lambda, \tau) \in R^N, \, N = N_1 + N_2, \begin{cases} v_i = \lambda_i & \text{for} 1 \leq i \leq N_1 \\ v_{j+N_1} = \tau_j & \text{for } 1 \leq j \leq N_2 \end{cases}.$$

Let $y(v)$ denote the solution to (6.1.2), (6.1.3) and (6.1.4) and C the linear operator "observation" of state y

$$z = Cy(v)$$

$$C: y \rightarrow z = (z_1, z_2, ..., z_M) \tag{6.1.8}$$

$z_k(t)$ is the pollutant concentration $y(x_k, t)$ measured at point x_k and at instant t. We have $z_k \in L^2(]0,T[)$ and $z \in H = L^2(]0,T[\times R^M)$. We call B the nonlinear operator that associates with parameters $v \in R^N$ the observation $Cy(v) \in H$:

$$B(v) = Cy(v) \tag{6.1.9}$$

We do not know the function $s : t \rightarrow s(t)$, or the function $y^0: x \rightarrow y^0(x)$, that is to say neither the pollution parameters λ_i nor the missing parameters τ_j. On the other hand, measurements z. are available.

The problem is to estimate the parameters $\{v_n\}_{1 \leq n \leq N_1}$, by using noisy measurements

$z_k(t)$ of $y(x_k, t)$, $1 \leq k \leq M$, $0 \leq t \leq T$

6.2 Sentinels of the linearized problem

Let $y(v)$ be the solution of the problem

$$
\begin{cases}
\dfrac{\partial y}{\partial t} - \Delta y + F(y) = \displaystyle\sum_{i=1}^{N_1} \lambda_i s_i(t)\delta(x-a) \text{ on } Q = \Omega \times]0,T[\\[4mm]
\dfrac{\partial y}{\partial n} = 0 \qquad\qquad\qquad\qquad\quad \text{on } \Sigma = \Gamma \times]0,T[\\[4mm]
y(x,0) = \displaystyle\sum_{j=1}^{N_2} \tau_j \chi_j(x) \qquad\qquad\quad \text{on } \Omega
\end{cases}
\tag{6.2.1}
$$

By linearizing equations (6.2.1) around $\{v, y(v)\}$ we obtain the system:

$$
\begin{cases}
\dfrac{\partial \hat{y}}{\partial t} - \Delta \hat{y} + F'(y)\hat{y} = \displaystyle\sum_{i=1}^{N_1} \hat{\lambda}_i s_i(t)\delta(x-a) \text{ on } Q = \Omega \times]0,T[\\[4mm]
\dfrac{\partial \hat{y}}{\partial n} = 0 \qquad\qquad\qquad\qquad\quad \text{on } \Sigma = \Gamma \times]0,T[\\[4mm]
\hat{y}(x,0) = \displaystyle\sum_{j=1}^{N_2} \hat{\tau}_j \chi_j(x) \qquad\qquad\quad \text{on } \Omega
\end{cases}
\tag{6.2.2}
$$

Equations in (6.2.2) are obtained by replacing in (6.2.1) y by $y + \hat{y}$ and v by $\hat{v}$, then keeping first-order terms only and dropping higher order terms.

Equations (6.2.2) define the differential at point v of the application y: $v \rightarrow y(v)$. It is the linear application $dy: \hat{v} \rightarrow \hat{y}$ associating $\hat{y}$ with $\hat{v} = \left(\hat{\lambda}, \hat{\tau}\right)$.

It is defined at point (v, y(v)) Let $B'(v): V \rightarrow H$ denote the differential application of B at v that associates z;^ with v;^. Provided this tangent linear application $B'(v): \hat{v} \rightarrow \hat{z}$ is injective, we can attach to it a generalized inverse, that is to say a family $W(v) = \{w^i(v), 1 \le i \le N\}$ of N sentinels w^i as we did in the above chapters for injective linear operators B. By convention the notation

$$
\hat{v} = W(v)\hat{z}
\tag{6.2.3}
$$

means that $\hat{y}$ is the column vector whose, for every index i, $1 \le i \le N$, i-th component is $\hat{v}_i = \left(w^i(v), \hat{z}\right)_H$, where $w^i(v)$ is the sentinel attached to the i-th parameter $\hat{v}_i (1 \le i \le N)$. We know that this generalized inverse $W(v)$ is the left inverse of $B'(v)$.

$$W(v)\,B'(v) = \text{Id}_V \qquad (6.2.4)$$

Our problem is to solve, in the mean squares sense, the equation

$$B(v) = z_d \qquad (6.2.5)$$

that is to say that we define a cost function

$$J(v) = \frac{1}{2}\,|B(v) - z_d|_H^2,\, v \in V$$

where $B(v) = Cy(v)$ is the observation of the nonlinear system *calculated* by solving (6.2.1) for $v = (\lambda, \tau)$ and where z_d is the *measured* concentration.

We are looking for $u \in V$ such that $J(u) \le J(v)\ \forall v \in V$.

Remark 6.2.1

When the observation is not exact there is no reason for equation (6.2.5) to admit a solution. It is the general case, because the observation of the state is "polluted" by noise and only provides an approximation z_d of Cy, which has no reason to be in $B(V)$.

6.3 Building the generalized inverse

Let $V = R^N$ be the space of parameters and H the space of observations. Let J be the cost function

$$J(v) = \frac{1}{2}\,|B(v) - z_d|_H^2,\, v \in V. \qquad (6.3.1)$$

Its derivative $J'(v)$ is such that

$$J'(v)w = (B(v) - z_d, B'(v)w)_H = (B'(v)^*(B(v) - z_d), w)_V,$$

and is given by

$$J'(v) = B'(v)^*(B(v) - z_d) \qquad (6.3.2)$$

If u minimizes J,

$$\begin{cases} u \in R^N \\ B'(u)^*(B(u) - z_d) = 0 \end{cases} \qquad (6.3.3)$$

We adopt, to minimize the cost function J, the iterative method

$$\begin{cases} v^0 \text{ initial guess} \\ p^n = -W^n(B^n - z_d) \\ v^{n+1} - v^n = p^n \end{cases} \tag{6.3.4}$$

where $B^n = B(v^n)$ and W^n is the generalized inverse of the linearized operator $B'(v^n)$

The linearized operator of B is $B'^n = B'(v_n)$, and the generalized inverse of B'^n is

$$W^n = \left(\Lambda^n\right)^{-1} B'^{n*} \tag{6.3.5}$$

where. $\Lambda^n = B'^{n*} B'^n$.

As in the preceeding chapters one easily checks that p^n is a descent direction.

6.4 Example

Let us return to the precise case we chose as an example in Section 6.1, the identification of parameters λ defining the function

$$s(t) = \sum_{i=1}^{i=N_1} \lambda_i s_i(t) \tag{6.4.1}$$

and (less importantly and maybe with difficulty) the identification of parameters τ defining the initial conditions

$$y^0(x) = \sum_{j=1}^{j=N_2} \tau_j \chi_j(x), x \in \Omega \tag{6.4.2}$$

We may be faced with two possible situations: in the first one we know the initial conditions, $y^0(x) = 0$, for example. In the second case we do not know them. But in all cases we need them to calculate the state $y(v)$, $v = (\lambda, \tau)$.

Space V

The vector v of parameters belongs to $V = R^N$ with
• in the first case $N = N_1$, the components of v being those of λ: $v_i = \lambda_i$ for $1 \leq i \leq N$

•in the second case $N = N_1 + N_2$, the components of v being in the order those of λ then those of τ

$$\begin{cases} v_i = \lambda_i & \text{for} \ 1 \le i \le N_1 \\ v_{j+N_1} = \tau_j & \text{for} \ 1 \le j \le N_2 \end{cases} \tag{6.4.3}$$

Space H

$H = L^2(]0,T[\times R^M)$, endowed with the usual inner product.

Non linear operator B

Given $v = (\lambda, \tau)$, find $y(v)$, solution of the nonlinear problem

$$\begin{cases} \dfrac{\partial y}{\partial t} - \Delta y + F(y) = \sum_{i=1}^{N_1} \lambda_i s_i(t)\delta(x-a) \ \text{on} \ Q = \Omega \times]0,T[\\[2mm] \dfrac{\partial y}{\partial n} = 0 \ \text{on} \ \Sigma = \Gamma \times]0,T[\\[2mm] y(x,0) = \sum_{j=1}^{N_2} \tau_j \chi_j(x) \ \text{on} \ \Omega \end{cases} \tag{6.4.4}$$

then $B(v) = z(v) = Cy(v) \in H$. $B(v)$ consists of M functions of t, $y(x_k, t, v)$, $1 \le k \le M$.

The observation $z_d \in H$, also consists of M functions of t, $z_{d, k}(t)$, $1 \le k \le M$.

Operator B'(v)

It is defined by

$$\begin{cases} \dfrac{\partial \hat{y}}{\partial t} - \Delta \hat{y} + F'(y)\hat{y} = \sum_{i=1}^{N_1} \hat{\lambda}_i s_i(t)\delta(x-a) \ \text{on} \ Q = \Omega \times]0,T[\\[2mm] \dfrac{\partial \hat{y}}{\partial n} = 0 \ \text{on} \ \Sigma = \Gamma \times]0,T[\\[2mm] \hat{y}(x,0) = \sum_{j=1}^{N_2} \hat{\tau}_j \chi_j(x) \ \text{on} \ \Omega \end{cases}$$

$$\tag{6.4.5}$$

$$C\Psi = [C\psi_1, \cdots, C\psi_N]$$

$$C\Psi(t) = [C\psi_1(t), \cdots, C\psi_N(t)] \in H$$

where ψ_i is the solution of (6.4.5) when all the components of $\hat{v}$ are zero, except the i-th one.

Let us denote

$$C\Psi = [C\Psi_1 \, C\Psi_2 \, ... \, C\Psi_N] \quad C\Psi(t) = [C\psi_1(t), \cdots, C\psi_N(t)] \tag{6.4.6}$$

An algorithm to identify parameters v for our nonlinear problem then is the following:

6.4.1 .Algorithm: Generalized inverse (nonlinear case)

initialize v by $v = v^0$ v^0 "arbitrary, or "at best" given the information we have about the parameters

relative distance d large, 10, 000, for example

while relative distance $> \varepsilon$, v being known, calculate the state $y = y(v)$ and the observation of this state $B(v) = Cy(v)$

knowing $(v, y(v))$ calculate the sentinels of the system linearized about this point:

$W = \{w^i(v), 1 \le i \le N\}$. Employing the direct method for example,

calculate

$$C\Psi(t) = [C\psi_1(t), \cdots, C\psi_N(t)]$$

calculate $\Lambda = \int_0^T C\Psi(t)^* C\Psi(t) dt$

solve for α $\Lambda\alpha = \text{Id}_V$

calculate $W = \Psi(t)\alpha$

calculate $p = -W(B - z_d)$

$v = v + p$

relative distance between two successive iterates

$$d = \frac{|p|}{|v|}$$

We see the parallelism between the linear and nonlinear cases. In each case we are faced with a homeomorphism B: $V \to B(V)$, B(V) closed in H. In both cases with each $z_d \in H$ we associate an element u of V that minimizes the distance between z_d and B(V).

Remark 6.4.1

In plus we have the state y.

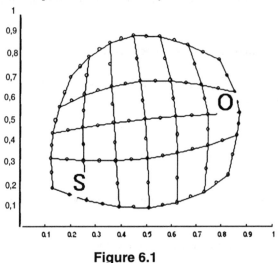

Figure 6.1

Spatial domain and a finite elements mesh, position of the source (S) and that of the observer (O). We divide the time interval into eight equal intervals; therefore there are 8 + 1 = 9 interval extremities and the same number of hat functions and parameters λ_i to identify.

Table 6.1

Norm of sentinels, indicating their sensibilities to noise

0.2449 0.1019 0.0809 0.0773   0.0769 0.0775 0.0802 0.0886 0.1266

Detail of the six iterations necessary to practically arrive at convergence

1.0000	2.0000	3.0000	4.0000	5.0000	6.0000	Iteration number
3.5497	2.8650	0.1201	0.0252	0.0006	0.0000	Distance between two successive iterates
38.9558	434.8101	10.5042	3.3677	3.2918	3.2918	cost function J

Evolution of parameters

0.0434	0.1429	0.1490	0.1492	0.1492	0.1492	Parameter 1
0.7921	0.4345	0.4195	0.4194	0.4194	0.4194	Parameter 2
1.4838	0.4486	0.4283	0.4282	0.4282	0.4282	Parameter 3
1.8986	0.3893	0.3686	0.3684	0.3684	0.3684	Parameter 4
1.8581	0.2256	0.2028	0.2028	0.2028	0.2028	Parameter 5
1.3078	0.0856	0.0542	0.0532	0.0532	0.0532	Parameter 6
0.7652	0.0768	0.0308	0.0294	0.0295	0.0295	Parameter 7
0.3081	0.1204	0.0479	0.0321	0.0318	0.0318	Parameter 8
0.5668	0.3371	0.2703	0.2507	0.2503	0.2503	Parameter 9

Figure 6.2

Sentinel attached to the first parameter. The time interval]0, 24[has been divided in to eight equal intervals ; therefore there are 8 + 1 = 9 interval ends and the same number of hat functions of sentinels and of parameters λ_i to identify.

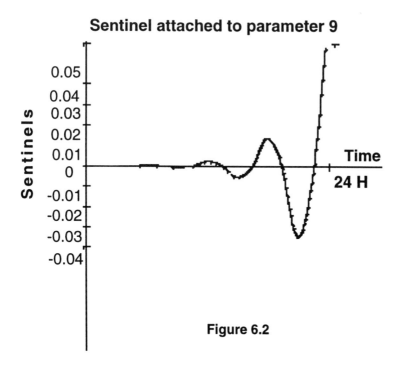

Figure 6.2

7 Dispersion coefficients

7.1. Motivation

The problem is to determine a dispersion coefficient for each element of a domain Ω (Figures 7.1 and 7.2) by using the measurement of the pollutant concentration at each of M observation points $x_1, x_2, \cdots, x_M$. Figure 7.1 shows that, in absence of noise, the recovered parameters are exact and figure 7.2 shows that, in presence of a noise of 20%, the recovered values of the dispersion parameter are very close to the exact values.

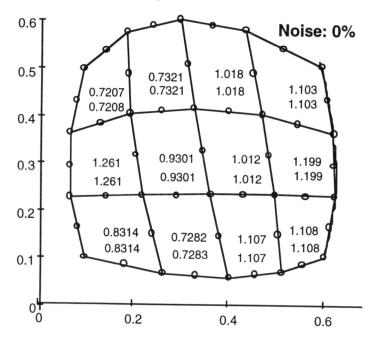

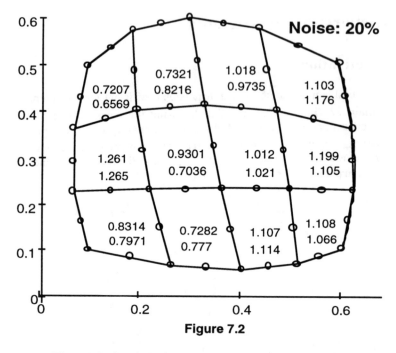

Figure 7.2

The sentinels method extends easily to this nonlinear situation. We show how simple it is to implement the method and how efficient it is. We work out the method for a dispersing and reacting pollutant. This method clearly applies to many possible inverse problems. In particular, the method could apply to the identification of the hydraulic conductivity in underground aquifers. It could also apply to electrical impedance tomography, where the problem is to find a conductivity distribution in a domain from electrostatic measurements collected at the boundary.

7.1.1 State of the system

An open set $\Omega \subset R^n$, n = 1, 2, or 3, is divided into N elements (a finite elements mesh as in Figure 1, for example). We suppose that the coefficient of dispersion is constant by element, of value λ_i on element $\Omega_i{}^1$. Let λ denote the vector of dispersion coefficients. $(\lambda_1, ..., \lambda_j, ..., \lambda_N)$.

The pollutant concentration at point x and time t, y = y(x, t), satisfies the state equation.

---

[1] Indeed, this is just a finite dimensional approximation of the actual dispersion coefficient on Ω.

$$y' + A(\lambda)y = f, \tag{7.1.1}$$

where

$$y' = \frac{\partial y}{\partial t}$$

and $A(\lambda)$ is the second-order elliptic operator

$$A(\lambda)y = -\mathrm{div}(\chi(\lambda)\mathrm{grad}\ y) + \sigma y$$

The parameter σ characterizes a first-order reaction consuming the pollutant. $\chi(\lambda)$ is the piecewise constant function:

$$\chi(\lambda) : x \to \sum_{i=1}^{N} \lambda_i \chi_i(x), \quad \chi_i \text{ is the characteristic function of the i-th}$$

element Ω_i , and

λ_i is the value of the function $\chi(\lambda)$ on element Ω_i, $1 \le i \le N$.1

$$\chi_i(x) = \begin{cases} 1 & \text{for } x \in \Omega_i \\ 0 & \text{otherwise} \end{cases}$$

$$\frac{\partial y}{\partial \nu} = 0 \text{ on } \Sigma = \Gamma \times (0,T) \tag{7.1.2}$$

where

$$\frac{\partial y}{\partial \nu_A} = \chi(\lambda)\mathrm{grad}\ \bar{y} \bullet \bar{n}, \quad \bar{n} \text{ being the unit outward normal.}$$

Suppose a null initial concentration of pollutant:

$$y(x,0) = 0 \tag{7.1.3}$$

and source terms,

$$f(x,t) = \sum_{i=1}^{i=N} s_i(t)\delta(x - a_i) \tag{7.1.4}$$

The N points a_i are given, together with the N functions s_i.

On the other hand, the parameter values λ_i are not given: they are the numbers we are looking for. If we are given these dispersion coefficients $\lambda = (\lambda_1, ..., \lambda_i, ..., \lambda_N)$, we can calculate the state y of the system by (7.1.1) together with boundary condition (7.1.2) and initial condition (7.1.3). Let $y(\lambda)$: $(x, t) \to y(x, t; \lambda)$ denote the state of the system corresponding to the dispersion coefficients λ.

7.1.2 Observation of the system

Suppose we only know approximate values of these parameters, say $\tilde{\lambda} = \left(\tilde{\lambda}_1, ..., \tilde{\lambda}_N \right)$. On the other hand, we have experimental data: the observed time history, as time t varies in the interval $]0, T[$, of the pollutant concentration $y\left(x_k, t \right)$ at M points x_k. Let C denote the exact observer: this is the observer providing exact measurements $z = B(v) = Cy(\lambda)$.

$$Cy(x, t; \lambda) = \{y(x_1, t; \lambda), ..., y(x_M, t; \lambda)\} \tag{7.1.5}$$

The actual observation z_d consists of M functions of time, $z_{d,k} : t \to z_{d,k}(t)$, which are noisy measurements of $y(x_k, t), k = 1, ..., M$.

$$z_d = Cy + \text{noise} \tag{7.1.6}$$

That is, let $z_{d,k} : t \to z_{d,k}(t)$ be the time history of the pollutant concentration at sensor k on the time interval (0,T), then the data corresponding to a source term f are

$$z_d = \left\{ z_{d,1}, ..., z_{d,M} \right\}. \tag{7.1.7}$$

Let us denote

$$H = L^2\left(0, T; R^M \right) \tag{7.1.8}$$

An identification experiment consists of generating a source term f of the form (7.1.4) and collecting the corresponding concentrations z_d. The problem is to improve our estimation of $\tilde{\lambda}$ by using the information hidden in z_d.

7.2 Linearized system

Together with the above quoted initial and boundary conditions the following equations define the states $\tilde{y} = y(\tilde{\lambda})$ and $y = y(\lambda)$

$$\tilde{y}' + A\left(\tilde{\lambda}\right)\tilde{y} = f \tag{7.2.1}$$

$$y' + A(\lambda)y = f \tag{7.2.2}$$

Here y and $\tilde{y}$ are the states respectively corresponding to the vectors of parameters λ and $\tilde{\lambda}$. Since $\tilde{\lambda}$ is known, $\tilde{y}$ can be calculated by using (7.2.1). But λ is unknown. However the principal value of the difference $\lambda - \tilde{\lambda}$, and hence an approximation of λ can be found by solving the inverse problem for the linearized problem. To get this linearized problem we substract (7.2.1) from (7.2.2), obtaining

$$\left(y - \tilde{y}\right)' + A(\lambda)y - A\left(\tilde{\lambda}\right)\tilde{y} = 0. \tag{7.2.3}$$

But

$$A(\lambda)y - A\left(\tilde{\lambda}\right)\tilde{y} = A\left(\tilde{\lambda}\right)\left(y - \tilde{y}\right) + A\left(\lambda - \tilde{\lambda}\right)\left(y - \tilde{y}\right) + A\left(\lambda - \tilde{\lambda}\right)\tilde{y} .$$

Dropping the term $\left[A\left(\lambda - \tilde{\lambda}\right)\right]\left(y - \tilde{y}\right)$. which is second order, we obtain the linearized equation

$$\eta' + A\left(\tilde{\lambda}\right)\eta = -A(\mu)\tilde{y} \quad (\text{+ boundary and initial conditions}) \tag{7.2.4}$$

where

$$\mu = \lambda - \tilde{\lambda}. \tag{7.2.5}$$

The solution η of the above equation (7.2.4) is the principal part of the difference between the states $y(\lambda)$ and $y\left(\tilde{\lambda}\right)$. The corresponding difference between the exact observation of y and that of $\tilde{y}$,

$$z = Cy(\lambda) \quad \text{and} \quad \tilde{z} = Cy\left(\tilde{\lambda}\right),$$

respectively is

$\varsigma = C\eta$, where $\varsigma = C\eta$ is a family of the M functions of time $\varsigma_k(t) = \eta(x_k, t), \ 1 \le k \le M$

We define the operator $B':V \to H$ by $B'\mu = C\eta$ where η is defined by (7.2.4). We define the sentinels $W: H \to V$ of the linearized system (i.e. the left inverse W of operator B'). Assuming $\tilde{\lambda}-\lambda$ "small" we have $\mu = \lambda - \tilde{\lambda} \cong W(z-\tilde{z})$ and assuming z_d-z "small" we have $W(z-\tilde{z}) \cong W(z_d - \tilde{z})$. Hence we can expect that a better estimation of λ is given by

$$\lambda \cong \tilde{\lambda} - W\left(z(\tilde{\lambda}) - z_d\right) \tag{7.2.6}$$

Remark 2.1

This formula is reminiscent of the formula given in Chapter 5 to update the successive approximations v^k of the vector of parameter values:

$$v^{k+1} = v^k - W\left(z(v^k) - z_d\right) \tag{7.2.7}$$

Consequently, we are led to iterate the above procedure and apply formula (7.2.7) until convergence. This is developed in Section 4.

7.3 Linearized system sentinels

Let $B:V = R^N \to H = L^2\left(]0,T[;R^M\right)$ be the (non linear) application that to v associates z and $B'(v):\hat{v} \to \hat{z}$ its tangent linear application which we suppose to be injective The differential of B at v is the application

$$\hat{v} \to \hat{z} = C\hat{y} = B'(v)\hat{v}. \tag{7.3.1}$$

For the linear problem (7.2.4) the sentinels are N functions w^n in $L^2(]0,T[;R^M)$ such that

$$(w^n, C\eta)_H = \mu_n \quad 1 \le n \le N \tag{7.3.1}$$

where $(.,.)_H$ is the inner product in H :

$$(w^n, \hat{z})_{H} = \sum_{k=1}^{M} \int_0^T w_k^n(t)\hat{z}_k(t)dt \tag{7.3.3}$$

Provided the tangent linear application B' is injective, one can attach to it a family $W(v)$ of N sentinels

$$W(v) = \left\{ w^n(v) \right\}_{1 \le n \le N}$$

is the operator defined by the N sentinels of the linearized system in the following way:

$$\hat{v} = W(v)\hat{z} \qquad (7.3.4)$$

means that $\hat{v}$ is the column vector whose n-th entry is

$$\hat{v}_n = (w^n(v), \hat{z})_H$$

where $w^n(v)$ is the sentinel associated with the n-th parameter $\hat{v}_n$, $(1 \le n \le N)$. We know that $W^n(v)$ is the left inverse of $B'(v)$.

$W(v)$ exists if and only if the application $B'(v)$ is injective, that is, if and only if the restrictions to $\omega \times]0,T[$ of the functions Ψ_i are linearly independent. Here Ψ_i is defined for $1 \le i \le N$ by

$$\begin{cases} \Psi_i' + A(\lambda)\Psi_i = -A(\chi_i)y(v) \\ \Psi_i(0 = 0 \\ \dfrac{\partial \Psi_i}{\partial n} = 0 \end{cases} \qquad (7.3.5)$$

Algorithm 7.1 to calculate the sentinels $W(v)$ of the linearized system

1. Define the "columns" of $B'(v)$ by

$$B'(v)e^i = C\Psi_i, \quad 1 \le i \le N.$$

2. Define the $N \times N$ matrix Λ by

$$\Lambda_i^j = \sum_{k=1}^{k=M} \int_0^T \Psi_i(x_k,t)\Psi_j(x_k,t)dt, \quad 1 \le i,j \le N.$$

3. For each $n \in \{1, 2, ..., N\}$ solve the linear system

$$\Lambda \gamma^n = e^n$$

4. Define w^n by

$$w^n = B\gamma^n$$

7.4 Nonlinear problem

The problem is to minimize the distance between the observation z_d and $B(v)$, where $B(v) = Cy(v) = Cy(\lambda)$ is calculated by solving (7.1.1), (7.1.5) and (7.1.7) while z_d is given by sensors.

Remark 7.4.1

There is no reason (except if the observation is exact) for the observation to coincide with an element z of H of the form z = B(v) with $v \in V$. An algorithm for identification of the parameters in V (including the calculation of the state y) for a nonlinear problem then is the following:

Algorithm 7.2 to identify the parameter values in the nonlinear case

1. Start from an arbitrary vector v^0
2. For n = 0, 1, 2, ..., until convergence, v^n being known
3. Calculate the state $y(v^n)$ and the observation $z(v^n) = Cy(v^n)$
4. Knowing $\left(v^n, y(v^n)\right)$ calculate (by algorithm 7.1) the sentinels of the system linearized around this point:

$$W(v^n) = \left[w^1(v^n), \cdots, w^1(v^n)\right]^T$$

5. Update the vector of parameters v^n by the formula

$$v^{n+1} = v^n - W(v^n)(B(v^n) - z_d)$$

Hypothesis 7.1.

We suppose that y is governed by (7.4.2)

Hypothesis 4.2

Let $\psi_i = \psi_i(\lambda)$ ($1 \leq i \leq N$ be the solution of the following problem:

$$\left|\begin{array}{l} \psi_i' + A(e^i)\psi_i = f \\ \psi_i(0) = 0 \\ \dfrac{\partial \psi_i}{\partial v_A} = 0 \end{array}\right. \qquad (7.4.1)$$

We suppose that the N functions $C\psi_i(\lambda)$ are linearly independent.

Let the observatory $\omega = \{x_k, 1 \le k \le M\}$ consist of M points of measurement x_k.

Proposition 4.1

Under hypotheses 7.4.1 and 7.4.2, the sentinel functions w^n exist and the algorithm converges (provided the starting point v^0 is taken close enough to the "best point" u).

Proof

$$C\hat{y} = 0 \Rightarrow \hat{\lambda} = 0 \Leftrightarrow \text{the N functions } C\Psi_i \text{ are linearly independent.}$$

Remark 7.4.1

That implies $\hat{y} = 0$.

Remark 7.4.2

The present algorithm does not converge a priori. Consider, for example, the system described by

$$\begin{cases} y' - \lambda \Delta y + \sigma\, y = 0 \\ y(0) = 0 \end{cases} \qquad (7.4.2)$$

where λ and σ are scalars. The unique solution of (7.4.2) is y = 0. It does not depend upon the coefficient of diffusion! Therefore we do not have to wonder if the algorithm sometime fails to give results.

Remark 7.4.4

It is easy to verify that hypothesis 7.4.2 is not satisfied in the case of equation (7.4.2) of remark 7.4.2. In fact in that case there is only one function ψ, defined by $\psi' - \lambda\Delta\psi = -div(\,grad\,y) = 0$ since y = 0, namely $\psi = 0$: the condition $\lambda\,C\psi = 0 \Rightarrow \lambda = 0$ is not satisfied.

7.5 Numerical results

7.5.1 First numerical experiment

Figure 7.3 shows the experimental set up (known positions of sources and sensors). Figures 7.4 to 7.7 show the estimated dispersion coefficients, under their exact values, for various noise levels.

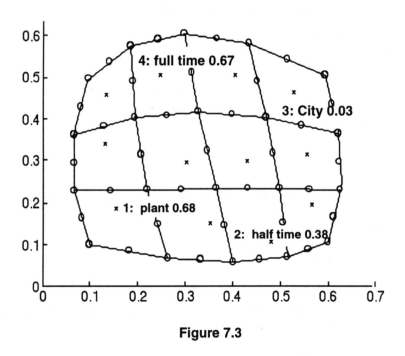

Figure 7.3

The spatial domain and its finite element mesh, the four source positions (labels 1 to 4) and many measurement points (stars).

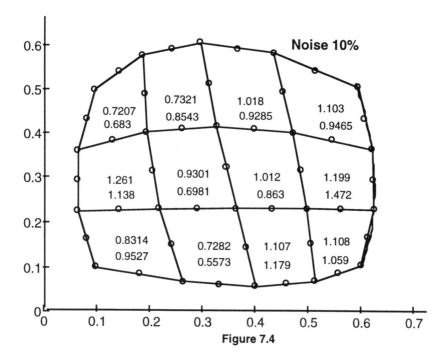

Figure 7.4

The dispersion coefficients restored in the presence of a 10% noise. In each element there are two numbers: the upper one is the exact value of the dispersion coefficient in this element, while the lower one is its restored value.

In the absence of noise, we recover the exact values, as shown in Figure 7.1.

Results for 10% noise are shown in Figure 7.2.

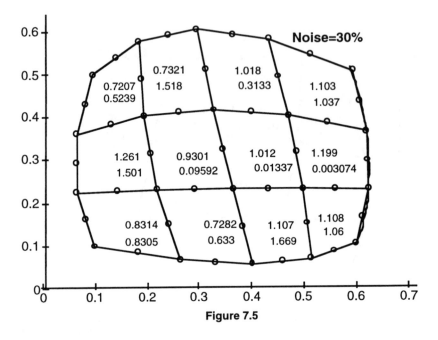

Figure 7.5

7.5.2 Second numerical experiment

Table 1

In each column, from top to bottom we find respectively the iteration number k, the distance from v^k to v^{k+1}, the value of the cost function $J(v^k)$, the norms of the sentinels, and the estimated parameter values v^k.

1.0000	2.0000	3.0000	4.0000	Iteration numbers
0.7051	0.1179	0.0064	0.0000	Variation between 2 successive v^k
0.2037	0.0393	0.0011	0.0000	Cost function
23.4899	16.5790	17.3986	17.4196	Norm of sentinels
41.7320	28.4249	30.2994	30.3741	
98.4770	71.5476	75.0771	75.2080	
55.5573	34.4268	37.4015	37.5387	
68.9694	44.0535	47.5938	47.7329	
51.8660	32.5607	35.5031	35.6410	
81.9468	58.4508	61.6815	61.8138	

18.0468	12.1946	12.9124	12.9400
43.5509	31.6524	33.2773	33.3629
96.1848	56.8983	63.1720	63.5307
34.1986	25.0507	26.2440	26.2974
68.6971	50.4414	52.8118	52.9003

0.9751	1.0079	1.0089	1.0089	Parameter values
1.1533	1.1457	1.1447	1.1446	
0.8446	0.9080	0.9109	0.9109	
1.1642	1.1497	1.1478	1.1478	
1.0106	1.0344	1.0347	1.0347	
0.8986	0.9528	0.9557	0.9557	
1.1284	1.1259	1.1248	1.1248	
0.9745	1.0090	1.0103	1.0103	
1.1845	1.1763	1.1760	1.1759	
0.9090	0.9641	0.9680	0.9680	
1.0045	1.0311	1.0319	1.0319	
1.0950	1.1004	1.0997	1.0997	

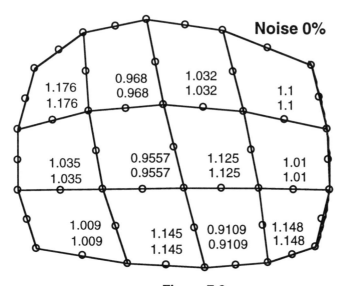

Figure 7.6

Table 7.2

1.0000	2.0000	3.0000	4.0000	Iteration number
0.6911	0.1126	0.0059	0.0001	Distance between 2 successive v^k
0.2040	0.0401	0.0086	0.0086	Cost function
23.4899	16.6726	17.4791	17.4975	Norm of sentinels
41.7320	28.9182	30.6829	30.7404	
98.4770	71.3440	74.8669	74.9982	
55.5573	35.4033	38.2958	38.4029	
68.9694	46.0557	49.1669	49.2462	
51.8660	33.5022	36.2340	36.3419	
81.9468	58.5961	61.7360	61.8579	
18.0468	12.2190	12.9670	12.9933	
43.5509	32.2905	33.7987	33.8683	
96.1848	57.8791	63.9287	64.2632	
34.1986	24.9745	26.1732	26.2325	
68.6971	50.5265	52.8336	52.9174	
0.9776	1.0096	1.0104	1.0104	Parameter values
1.1457	1.1398	1.1390	1.1390	
0.8646	0.9244	0.9267	0.9267	
1.1380	1.1305	1.1292	1.1292	
1.0321	1.0508	1.0506	1.0506	
0.9155	0.9655	0.9678	0.9678	
1.0982	1.1021	1.1018	1.1018	
0.9890	1.0197	1.0208	1.0208	
1.1929	1.1834	1.1828	1.1828	
0.8981	0.9557	0.9602	0.9602	
1.0077	1.0335	1.0340	1.0340	
1.0705	1.0810	1.0810	1.0810	

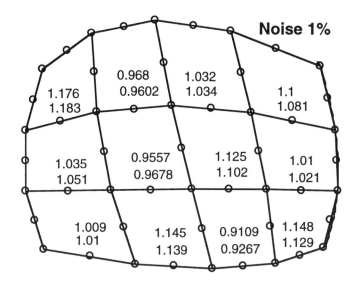

Figure 7.7

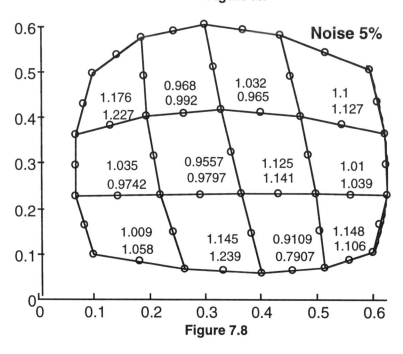

Figure 7.8

Table 7.3

1.0000	2.0000	3.0000	4.0000	5.0000	6.0000	7.0000	Iteration numbers
0.8198	0.1967	0.0316	0.0076	0.0018	0.0004	0.000	Distance between 2 successive v^k
0.2120	0.0648	0.0463	0.0462	0.0462	0.0462	0.0462	function
23.4899	17.2444	17.9048	17.9066	17.9119	17.9108	17.9111	Norm of sentinels
41.7320	27.6355	30.7849	30.5811	30.6594	30.6408	30.6453	
98.4770	79.3496	80.7654	81.9177	81.7808	81.8177	81.8089	
55.5573	29.3352	34.7520	33.7631	33.9760	33.9227	33.9356	
68.9694	38.8430	45.4660	45.1775	45.3569	45.3156	45.3258	
51.8660	31.2988	36.1873	35.7956	35.9281	35.8967	35.9043	
81.9468	64.2922	66.4407	67.0607	67.0102	67.0252	67.0217	
18.0468	13.0859	12.5705	12.7381	12.7048	12.7130	12.7110	
43.5509	29.7578	33.1274	32.7493	32.8527	32.8278	32.8339	
96.1848	62.0592	67.6407	68.1710	68.1667	68.1704	68.1696	
34.1986	31.5442	30.6701	31.2321	31.1422	31.1655	31.1599	
68.6971	56.5463	57.3274	57.9557	57.8828	57.9028	57.8980	
1.0431	1.0585	1.0584	1.0584	1.0584	1.0584	1.0584	Parameter values
1.2832	1.2381	1.2398	1.2385	1.2388	1.2387	1.2387	
0.6602	0.8022	0.7874	0.7915	0.7905	0.7908	0.7907	
1.1354	1.1093	1.1067	1.1056	1.1058	1.1057	1.1057	
0.9071	0.9754	0.9728	0.9746	0.9742	0.9743	0.9742	
0.9128	0.9821	0.9782	0.9801	0.9797	0.9798	0.9797	
1.1672	1.1280	1.1440	1.1406	1.1414	1.1412	1.1412	
1.0031	1.0353	1.0388	1.0389	1.0389	1.0389	1.0389	
1.2287	1.2299	1.2259	1.2271	1.2269	1.2269	1.2269	
0.9678	0.9797	0.9944	0.9914	0.9921	0.9920	0.9920	
0.9194	0.9705	0.9640	0.9653	0.9650	0.9651	0.9650	
1.1424	1.1146	1.1291	1.1262	1.1269	1.1267	1.1268	

Note: We see that already the first iteration yields a value close to the final one.

8 Position of a source

Let us consider a two-dimensional region Ω occupied by polluted water. In the preceeding chapters we knew the position of each source of pollution . In this chapter we no longer know it. However, to the question whether it is possible to identify this position, the answer is yes!

In Section 8.1 we pose the problem for a single source, located at an unknown fixed point a and discharging a pollutant at a rate that is a known function of time. The problem is to find the point (+), by using the measurement of the pollutant concentration on the time interval [0, T] at M points of observation, denoted O in Figure 8.1:

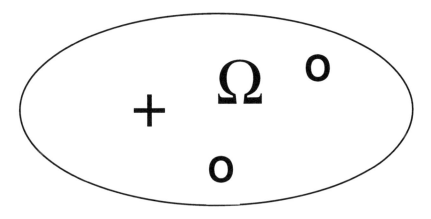

Figure 8.1

We are going, as in the previous chapters, to successively examine the following steps:

1. Linearization   (Section 8.2)

2. Computation of the linearized system sentinels   (Section 8.3)

3. Use of these sentinels to find the point a   (Section 8.4)

4. Some identification experiments   (Section 8.5)

The cases when this position not only is unknown but moreover varies as an unknown function of time are also considered.

In Section 8.6, both the position a and the time history $s = s(t)$ of the source are unknown.

In Section 8.7, we consider the case of a moving source, $a = a(t)$, whose trajectory, $a = a(t)$, is unknown, together with the time history $s = s(t)$.

We also could consider the case of a moving observatory: $x_k = x_k(t)$ (k = 1, ..., M). Of course in general the functions $x_k(t)$ are known.

8.1 Position of the problem

The state y is solution of the equation

$$\frac{\partial y}{\partial t} - \Delta y + \sigma y = s(t)\delta_a \quad \text{on } Q = \Omega \times]0, T[\qquad (8.1.1)$$

where $y = y(x, t)$ is the pollutant concentration at point $x = (x_1, x_2)$ and at instant t, s(t) is the flow rate, which we suppose to be known, of a single source of pollutant located at point $a = (a_1, a_2)$, which we suppose to be unknown, δ_a is the Dirac mass at point a, that is, the distribution defined by

$$\langle \delta_a | \varphi \rangle = \varphi(a) \; \forall \varphi \in \mathcal{D}(\Omega)$$

which is sometimes denoted (improperly because the Dirac mass is not a function)

$$\int_\Omega \delta(x-a)\varphi(x)dx \text{ where } x = (x_1, x_2) \text{ and } dx = dx_1 dx_2$$

and σ is a reaction coefficient (disappearance of the pollutant in a reaction of velocity σy).

The boundary condition is a zero flux condition, expressing that the pollutant cannot flow through the boundary of Ω:

$$\frac{\partial y}{\partial n} = 0 \quad \text{on} \quad \Sigma = \Gamma \times] 0, T[\qquad (8.1.2)$$

where $\dfrac{\partial y}{\partial n} = \vec{n} \bullet \nabla y$ is the unit outward normal derivative, $\vec{n}$ being the unit outward normal.

The initial condition is

$y(x, 0) = y_0(x)$ on Ω, $\qquad\qquad\qquad\qquad (8.1.3)$

which we suppose to be known, $y_0(x) = 0$, for example.
    With our familiar notations we have

$$\begin{cases} y' + Ay = s\delta_a \\ y(0) = 0 \end{cases} \qquad\qquad\qquad\qquad (8.1.4)$$

where the operator $A = -\Delta + \sigma$ is operating on functions y satisfying the boundary condition (8.1.4). Let Y denote the set of functions y satisfying (8.1.4) as the function s spans $L^2(0, T)$ and the point a takes all positions in Ω, and $C: Y \to H$ be the linear operator that gives the exact observation z of state y: $Cy = z = (z_1, z_2,..., z_M)$ where $z_k(t) = y(x_k,t)$, pollutant concentration at point x_k and at instant t. To identify coordinates (a_1, a_2) of point a, we have at our disposal a noisy observation z_d of the state y.

$z_{d,k}(t) = y(x_k, t) + \text{noise}(k, t)$.

    The problem is to identify the two parameters (a_1, a_2) knowing z_d. Here the (nonlinear) application B is from R^2 into $H = L^2(0, T; R^2)$.

8.2. Linearization

    We rewrite equations (8.1.4) with $a + \hat{a}$ in place of a,

$$\begin{cases} (y + \hat{y})' + A(y + \hat{y}) = s(\delta_{a+\hat{a}}) \\ (y + \hat{y})(0) = 0 \end{cases}$$

then substract, thus obtaining

$$\begin{cases} \hat{y}' + A\hat{y} = s\delta_{a+\hat{a}} - s\delta_a \\ \hat{y}(0) = 0 \end{cases}$$

The right hand side is the distribution defined by

$$\langle \delta_{a+\hat{a}} - \delta_a | \varphi \rangle = \varphi(a + \hat{a}) - \varphi(a) = \langle \nabla\varphi(a) | \hat{a} \rangle + \text{h.o.t.}$$

$$= \frac{\partial \varphi}{\partial x_1}(a)\hat{a}_1 + \frac{\partial \varphi}{\partial x_2}(a)\hat{a}_2 + h.o.t. \; \forall \varphi \in D(\Omega)$$

But the derivative δ'_a of δ_a is precisely such that

$$\langle \delta'_a \hat{a} | \varphi \rangle = -\langle \nabla \varphi(a) | \hat{a} \rangle \; \forall \varphi \in D(\Omega)$$

In conclusion, dropping the h.o.t.s, we are led to define the linearized problem as the one which assocates to a deviation $\hat{a}$ such that $a + \hat{a} \in \Omega$ a corresponding variation $\hat{y}$ of y defined by (8.2.1).

$$\begin{cases} \hat{y}' + A\hat{y} = s\delta'_a \hat{a} \\ \hat{y}(0) = 0 \end{cases} \tag{8.2.1}$$

Another way to say things is the following: Let y(a): (x, t) → y(x, t; a) be the solution of problem (8.1.1), (8.1.2), and (8.1.3). We define the differential $y'(a): \hat{a} \to \hat{y}$ of y at a by (8.2.1) where

$$\langle \delta'_a \hat{a} | \varphi \rangle = -\frac{\partial \varphi(a)}{\partial x_1}\hat{a}_1 - \frac{\partial \varphi(a)}{\partial x_2}\hat{a}_2 = -\langle \nabla \varphi(a) | \hat{a} \rangle \tag{8.2.2}$$

Thus, "morally", $\hat{z} = C\hat{y}$ is the variation of the exact observation corresponding to the variation $\hat{a} = (\hat{a}_1, \hat{a}_2)$ of the position of the source point a.

8.3 Sentinels of the linearized problem

For this linearized problem we can solve the sentinels problem: find two functions w^i of H such that

$$\left(w^i, \hat{z}\right)_H = \hat{a}_i, \; 1 \le i \le N$$

where $\hat{z} = C\hat{y}$ is the observation of $\hat{y}$, $C\hat{y}(t) = \left(\hat{y}(x_1, t), \; ..., \; \hat{y}(x_M, t)\right)$ and $\hat{z}_k(t) = \hat{y}(x_k, t)$.

$$\left(w^i, \; \hat{z}\right)_H = \sum_{k=1}^{M} \int_{]0,T[} w^i_k(t)\hat{z}_k(t)dt,$$

inner product in $L^2(]0, T[; R^M)$.

We thus define two sentinels for the linearized system, the former, w_1, to identify $\hat{a}_1$, the latter, w_2, to identify $\hat{a}_2$ the observation being $\hat{z}$. The knowledge of these two sentinels is interesting because it enables us to infer from a "small" discharge $\hat{z}$ of z what "small" deviation $\hat{z}$ of a is the cause of $\hat{z}$, and to update the state of the system: update the position of the source to $a + \hat{a}$.

Given a point a, for $i = 1$ and for $i=2$ $w^i(a)$ is obtained, either by the direct method or by the indirect method. Let us make more precise these two methods to calculate $W(a) = \begin{bmatrix} w^1(a) \\ w^2(a) \end{bmatrix}$.

8.4. Nonlinear problem

The problem to solve is

$$\inf_{a \in \Omega \cup \Gamma} |B(a) - z_d|_H \qquad (8.4.1)$$

where z_d denotes the (experimental) observation of y. As in chapters 1, 2, and 5 we can employ the sentinels of the linearized problerm:

8.4.1 Sentinels algorithm (non linear case)

An iteration is described below. We stop the iterations when the relative distance between two successive iterates a^n and a^{n+1} is less than ε:

$$|a^{n+1} - a^n| \le \varepsilon |a^{n+1}|$$

a being given, solve for y the system

$$y' + Ay = s(t) \, \delta(x - a) \, ; \, y(0) = 0 \qquad (8.1.4)$$

restrict the calculated state y to the calculated observation z_c of the state y: $z_c = Cy = B(a)$
(c for calculated, whereas in z_d, d stands for desired)

calculate the sentinels W of the linearized system: $W = \begin{cases} w1 \\ w2 \end{cases}$

calculate the distance between two successive iterates: $e = W(a) \, (z_c - z_d)$
update a: $a = a - e$

8.5. Numerical experiments

The following numerical experiments show the convergence of this algorithm even for starting points a^0 far from the actual position of the source. The basin of attraction is large. With a single observer, according to the starting point, either there is convergence to a spurious solution, or at some iteration n the point a^n quits the domain, or there is convergence of a^n towards the good solution (the source position), herein labelled with the + sign: $a^n \rightarrow a$.

8.5.1 Numerical experiment 1

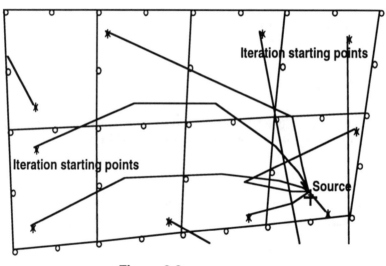

Figure 8.2

Some starting points generate a convergente sequence, others do not.

8.5.2 Numerical experiment 2

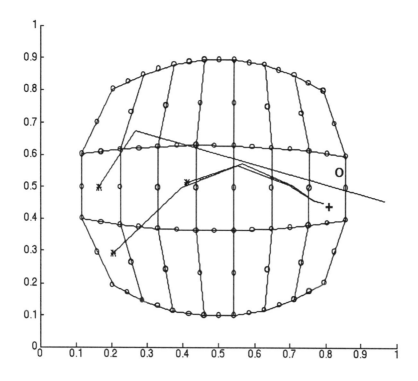

o, observer

+, source

*, iterations starting point

Figure 8.3

Figure 8.3 *shows the iteration steps according to the starting points.*

8.5.3 Numerical experiment 35.3

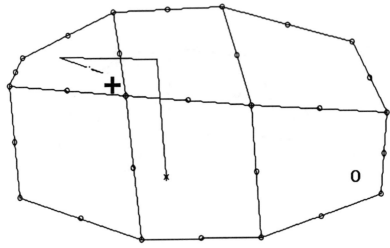

+: source
O: observer
*: start

Figure 8.4

8.6. Unknown position and flow rate

Governing equations

$$\begin{cases} \dfrac{\partial y}{\partial t} - \Delta y + \sigma y = s(t)\delta_a \\ \dfrac{\partial y}{\partial n} = 0 \\ y(x,0) = 0 \end{cases} \qquad (8.6.1)$$

Pollutant concentration modelling

The pollutant is discharged in sea by a source located at point $a \in \Omega$, $\Omega \subset R^2$. The pollutant flow rate is the function of time s: $]0, T[\rightarrow R : t \rightarrow s(t)$, called the time-history of the source.

What is unknown and has to be identified

Point $a = (a_1, a_2) \in R^2$ and function $s \in L^2(0, T)$.

What is known

The observation $z_k(t)$ of pollutant concentration at points x_k ($1 \leq k \leq M$) during the time interval $] 0, T[$). The observation $z = \{z_1, z_2, ..., z_M\} \in H = L^2(]0, T[, R^M)$

Approximation of function s

The time interval $(0,T)$ is divided into $N - 1$ sub-intervals (t_i, t_{i+1}) of length $\Delta t_i = t_{i+1} - t_i$. The "hat function" s_i: $t \rightarrow s_i(t)$ is the continuous function affine on each interval (t_j, t_{j+1}), $0 \leq j \leq N-1$, and $s(t_j) = \delta_i^j$, $0 \leq j \leq N$. We look for the unknown function s as a linear combination of such "hat functions", thus finding a piecewise affine continuous function.

$$s(t) = \sum_{i=1}^{i=N} \lambda_i s_i(t).$$
(8.6.2)

Thus the state y is defined by

$$y' + Ay = s\delta_a \quad y(0) = 0$$
(8.6.3)

where A is the already defined operator $A = - \Delta + \sigma$ acting on functions in $H^1(\Omega)$ satisfying a zero flux boundary condition and s is given by (8.6.2).

Equations of the linearized problem

$$\begin{cases} \dfrac{\partial \hat{y}}{\partial t} - \Delta \hat{y} + \sigma \hat{y} = \hat{s}(t)\delta_a + s(t)(-\delta_a' \hat{a}) \text{ in } Q \\ \dfrac{\partial \hat{y}}{\partial n} = 0 \text{ on } \Sigma \\ \hat{y}(x, 0) = 0 \text{ on } \Omega \end{cases}$$
(8.6.4)

where

$$\hat{s}(t) = \sum_{i=1}^{N} \hat{\lambda}_i s_i(t), \quad \hat{a} = (\hat{a}_1, \hat{a}_2), \quad \hat{a} = \begin{pmatrix} \hat{a}_1 \\ \hat{a}_2 \end{pmatrix}, \quad \hat{a}(t) = \begin{cases} \sum_{j=1}^{N} \hat{\mu}_{1,j} s_j(t) \\ \sum_{j=1}^{N} \hat{\mu}_{2,j} s_j(t) \end{cases} \tag{8.6.5}$$

$$\langle \delta'_a \hat{a} | \varphi \rangle = -\frac{\partial \varphi(a)}{\partial x_1} \hat{a}_1 - \frac{\partial \varphi(a)}{\partial x_2} \hat{a}_2 = -\langle \nabla \varphi(a) | \hat{a} \rangle \quad \forall \varphi \in D(\Omega) \tag{8.6.6}$$

How have equations been obtained? By first writing equations (8.6.2) with s and a of the form (8.6.3), then replacing s and a by $s + \hat{s}$ and $a + \hat{a}$ where s and a are of the form (8.6.5), then substracting:

$$(y + \hat{y})' + A(y + \hat{y}) = (s + \hat{s})(\delta_{a+\hat{a}}), \qquad (y + \hat{y})(0) = 0$$

$$y' + Ay = s\delta_a, \quad y(0) = 0$$

$$\hat{y}' + A\hat{y} = (s + \hat{s})(\delta_{a+\hat{a}}) - s\delta_a, \hat{y}(0) = 0$$

$$\langle \delta'_a \hat{a} | \varphi \rangle = -\frac{\partial \varphi(a)}{\partial x_1} \hat{a}_1 - \frac{\partial \varphi(a)}{\partial x_2} \hat{a}_2 = -\langle \nabla \varphi(a) | \hat{a} \rangle \quad \forall \varphi \in \mathbf{D}(\Omega)$$

$$\langle \delta_{a+\hat{a}} - \delta_a | \varphi \rangle = \varphi(a + \hat{a}) - \varphi(a) = (\nabla \varphi(a), \hat{a})_{R^2} + h.o.t. = \langle -\delta'_a \hat{a} | \varphi \rangle + h.o.t.$$

In conclusion, dropping the h.o.t.s, we are led to define the linearized problem

$$\hat{y}' + A\hat{y} = s\delta'_a \hat{a} + \hat{s}\delta_a , \quad \hat{y}(0) = 0 \tag{8.6.7}$$

The equations of the linearized system are (8.6.4), (8.6.5) , and (8.6.6).

8.7 Sentinels of the linearized problem (1)

In order to determine the sentinels of the linearized problem by the direct method, we need the "response" of the linearized system to each parameter, i.e. the observation $C\hat{y}$ when all the $\hat{\lambda}_i$, $1 \le i \le N$, $\hat{a}_1$, and $\hat{a}_2$ are

0, except one equal to 1. We thus obtain the functions $\psi_i(x,t), \varphi_1(x,t)$ and $\varphi_2(x,t)$. In order to calculate $\psi_i(x,t), \varphi_1(x,t)$ and $\varphi_2(x,t)$. we solve the problems:

$$\begin{cases} \dfrac{\partial \psi_i}{\partial t} - \Delta \psi_i + \sigma \psi_i = s_i(t)\delta_{a(t)} \ in \ Q \\ \dfrac{\partial \psi_i}{\partial n} = 0 \ on \ \Sigma \\ \psi_i(x,0) = 0 \ on \ \Omega \end{cases} \qquad (8.7.1)$$

To define φ_1 we take $\hat{\lambda}_j = 0 \forall j$, $\hat{a}_1 = 1$ and $\hat{a}_2 = 0$

$$\begin{cases} \dfrac{\partial \varphi_1}{\partial t} - \Delta \varphi_1 + \sigma \varphi_1 = s(t)\delta'_a \begin{bmatrix} 1 \\ 0 \end{bmatrix} \ in \ Q \\ \dfrac{\partial \varphi_1}{\partial n} = 0 \ on \ \Sigma \\ \varphi_1(x,0) = 0 \ on \ \Omega \end{cases} \qquad (8.7.2)$$

In fact, what we use to define φ_1 is the weak formulation

$$\frac{d}{dt}(\varphi_1, v) + a(\varphi_1, v) = s(t)\frac{\partial v(a)}{\partial x_1} \qquad \forall v \in H^1(\Omega)$$

For φ_2 we have the weak formulation

$$\frac{d}{dt}(\varphi_2, v) + a(\varphi_2, v) = s(t)\frac{\partial v(a)}{\partial x_2} \qquad \forall v \in H^1(\Omega)$$

Once we have calculated (by the finite element method, for example) the functions ψ_i, φ_1 , and φ_2 we have in particular the observations $C\psi_i$, $C\varphi_1$, and $C\varphi_2$, and we can build the matrix Λ of their inner products in H.

8.8 Sentinels of the linearized problem (2)

In order to determine the sentinels of the linearized problem we solve the linear-quadratic control problem whose pseudo state ρ is the solution of the system

$$\begin{cases} \rho' + A\rho = -s\delta'_a \alpha + \sigma\delta_a & \rho(0) = 0 \\ \sigma = \sum_{i=1}^{i=N} \sigma_i s_i(t), \sigma_i \in R, \alpha \in R^2 \end{cases} \qquad (8.8.1)$$

where α and σ are control vectors, defined by the N-dimensional vector $\{\sigma_i\}_{1\le i\le N}$, α_1, and α_2.

These parameters α and σ appear in the pseudo-state equations exactly like λ and in the linearized state equations. The cost function to be minimized in order to obtain the sentinel attached to α_n or σ_n is

$$J(\sigma, \alpha) = \frac{1}{2}|C\rho|^2_H - (\sigma_n \text{ or } \alpha_n) \qquad (8.8.2)$$

and the corresponding sentinel is given by the observation of ρ at optimality:

$$w = C\rho \qquad (8.8.3)$$

8.9 Moving source

In contrast to the preceeding case, the position of the boat that discharges the pollutant is moving with time.

Governing equations

$$\begin{cases} \dfrac{\partial y}{\partial t} - \Delta y + \sigma y = s(t)\delta_{a(t)} & \text{in } Q \\ \dfrac{\partial y}{\partial n} = 0 & \text{on } \Sigma \\ y(x, 0) = 0 & \text{on } \Omega \end{cases} \qquad (8.9.1)$$

Modelling the pollutant concentration y

The pollutant is discharged into the sea by a moving boat .The path followed by the boat is defined by the trajectory equation a: $]0, T[\to R^2 : t \to a(t)$: at time t the boat is at point a(t).

The pollutant flow rate is the function s of time s: $]0, T[\to R : t \to s(t)$.

What is unknown and has to be identified

Both functions a and s.

What is known

The observation $z_k(t)$ of the pollutant concentration at point x_k ($1 \leq k \leq M$) and at time t ($t \in] 0, T[$). The observation $z = \{z_1, z_2, ..., z_M\} \in H = L^2(]0, T[, R^M)$.

Approximation of functions s and a

The time interval (0,T) is divided into N-1 sub-intervals (t_i, t_{i+1}) of length $\Delta t_i = t_{i+1} - t_i$. The hat function $s_i: t \to s_i(t)$ is affine by interval (t_j, t_{j+1}), $0 \leq j \leq N-1$, and $s(t_j) = \delta_i^j$ for $0 \leq j \leq N$. We look for the unknown functions s and a = (a_1, a_2) as linear combinations of such hat functions. Thus the state y is defined by

$$y' + Ay \ \delta = s(t)\delta(x-a(t)) \tag{8.9.2}$$

where A is the already defined operator A = $-\Delta + \sigma$ acting on functions in $H^1(\Omega)$ satisfing a zero flux boundary condition

$$s(t) = \sum_{i=1}^{i=N} \lambda_i s_i(t), \ a(t) = \begin{pmatrix} a_1(t) \\ a_2(t) \end{pmatrix}, \ a(t) = \begin{cases} \sum_{j=1}^{N} \mu_{1,j} s_j(t) \\ \sum_{j=1}^{N} \mu_{2,j} s_j(t) \end{cases} \tag{8.9.3}$$

$$a(t) \in R^2. \tag{8.9.4}$$

Equations of the linearized problem

$$\left| \begin{array}{l} \dfrac{\partial \hat{y}}{\partial t} - \Delta \hat{y} + \sigma \hat{y} = \hat{s}(t)\delta_{a(t)} - s(t)\delta'_a \hat{a}(t) \ in \ Q \\[2mm] \dfrac{\partial \hat{y}}{\partial n} = 0 \ on \ \Sigma \\[2mm] \hat{y}(x,0) = 0 \ on \ \Omega \end{array} \right.$$

(8.9.5)

where

$$\hat{s}(t) = \sum_{i=1}^{N} \hat{\lambda}_i s_i(t), \quad \hat{a}(t) = \begin{pmatrix} \hat{a}_1(t) \\ \hat{a}_2(t) \end{pmatrix}, \quad \hat{a}(t) = \begin{cases} \sum_{j=1}^{N} \hat{\mu}_{1,j} s_j(t) \\ \sum_{j=1}^{N} \hat{\mu}_{2,j} s_j(t) \end{cases}$$

(8.9.6)

$$\langle \delta'_a \hat{a} | \varphi \rangle = -\frac{\partial \varphi(a)}{\partial x_1} \hat{a}_1 - \frac{\partial \varphi(a)}{\partial x_2} \hat{a}_2 = -\langle \nabla\varphi(a) | \hat{a} \rangle \ \forall \varphi \in \ D(\Omega)$$

(8.9.7)

How have these equations been obtained?

By first writing equations (8.9.2) with s and a of the form (8.9.3) and (8.9.4), then replacing s and a

by $s + \hat{s}$ and $a + \hat{a}$ where $\hat{a}(t) = \begin{pmatrix} \hat{a}_1(t) \\ \hat{a}_2(t) \end{pmatrix}$, then substracting

$$y' + Ay = s\delta_a \qquad\qquad\qquad y(0) = 0$$

$$\hat{y}' + A\hat{y} = (s + \hat{s})(\delta_{a+\hat{a}}) - s\delta_a \qquad\qquad \hat{y}(0) = 0$$

$$\hat{y}' + A\hat{y} = s(\delta_{a+\hat{a}} - \delta_a) + \hat{s}\delta_{a+\hat{a}} = s(\delta_{a+\hat{a}} - \delta_a) + \hat{s}\delta_a + \hat{s}(\delta_{a+\hat{a}} - \delta_a)$$

$$\langle \delta'_a \hat{a} | \varphi \rangle = -\frac{\partial \varphi(a)}{\partial x_1} \hat{a}_1 - \frac{\partial \varphi(a)}{\partial x_2} \hat{a}_2 = -\langle \nabla\varphi(a) | \hat{a} \rangle \ \forall \varphi \in \ D(\Omega)$$

$$\left\langle \delta_{a+\hat{a}} - \delta_a \middle| \varphi \right\rangle = \varphi(a+\hat{a}) - \varphi(a) = \left(\nabla\varphi(a), \hat{a}\right)_{R^2} + h.o.t.s = \left\langle -\delta'_a \hat{a} \middle| \varphi \right\rangle$$

In conclusion, dropping the h.o.t.s, we are led to define the linearized problem.(8.9.5) , (8.9.6), and (8.9.7).

Sentinels of the linearized problem

In order to determine the sentinels of the linearized problem by the direct method we need the "response" of the linearized system to each parameter, i.e., the observation $C\hat{y}$ when all the $\hat{\lambda}_i, \hat{\mu}_{1,i}, \hat{\mu}_{2,i}$ $1 \leq i \leq N$ are 0, except one, equal to 1. We thus obtain the functions $\psi_i(x,t), \varphi_{1,i}(x,t)$ and $\varphi_{2,i}(x,t)$. In order to calculate $\psi_i(x,t), \varphi_{1,i}(x,t)$ and $\varphi_{2,i}(x,t)$ we solve the problems:

$$\begin{cases} \dfrac{\partial\psi_i}{\partial t} - \Delta\psi_i + \sigma\psi_i = s_i(t)\delta_{a(t)} & \text{in } Q \\[2ex] \dfrac{\partial\psi_i}{\partial n} = 0 & \text{on } \Sigma \\[2ex] \psi_i(x,0) = 0 & \text{on } \Omega \end{cases} \qquad (8.9.10)$$

To define $\varphi_{1,i}$ we take $\hat{\lambda}_j = 0 \forall j$ $\qquad \hat{\mu}_{1,j} = \begin{cases} 1 \text{ if } j = i \\ 0 \text{ if } j \neq i \end{cases}$ $\qquad \hat{\mu}_{2,j} = 0 \forall j$

$$\begin{cases} \dfrac{\partial\varphi_{1,i}}{\partial t} - \Delta\varphi_{1,i} + \sigma\varphi_{1,i} = -s(t)s_i(t)\delta'_{a(t)}\begin{bmatrix} 1 \\ 0 \end{bmatrix} \\[2ex] \dfrac{\partial\varphi_{1,i}}{\partial n} = 0 \\[2ex] \varphi_{1,i}(x,0) = 0 \end{cases} \qquad (8.9.11)$$

In fact, what we use to define $\varphi_{1,i}$ is the weak formulation

$$\frac{d}{dt}\left(\varphi_{1,i}, v\right) + a\left(\varphi_{1,i}, v\right) = s(t)s_i(t)\left(-\frac{\partial v}{\partial x_1}(a(t))\right)$$

$$\begin{cases} \dfrac{\partial \varphi_{1,i}}{\partial t} - \Delta\varphi_{1,i} + \sigma\varphi_{1,i} = -s(t)s_i(t)\delta'_{a(t)}\begin{bmatrix} 1 \\ 0 \end{bmatrix} \\[2mm] \dfrac{\partial \varphi_{1,i}}{\partial n} = 0 \\[2mm] \varphi_{1,i}(x,0) = 0 \end{cases}$$

$$(8.9.11)$$

For $\varphi_{2,i}$ we have the weak formulation

$$\frac{d}{dt}\left(\Phi_{2,i},v\right) + a\left(\Phi_{2,i},v\right) = s(t)s_i(t)\left(-\frac{\partial v}{\partial x_2}(a(t))\right)$$

Once we have calculated the functions ψ_i, $\psi_{1,i}$ and $\varphi_{2,i}$ we have in particular the observations $C\psi_i$, $C\varphi_{1,i}$ and $C\varphi_{2,i}$ and we can build the matrix Λ of their inner products in H.

Sentinels of the linearized problem (4)

In order to determine the sentinels of the linearized problem we solve the linear-quadratic control problem whose pseudo state ρ is the solution of the system

$$\begin{cases} \rho' + A\rho = s\delta'_a\alpha + \sigma\delta_a \;, \quad \rho(0) = 0 \\[2mm] \sigma = \sum_{i=1}^{i=N}\sigma_i s_i(t), \sigma_i \in R, \alpha = \sum_{i=1}^{i=N}\alpha_i s_i(t), \alpha_i \in R^2 \end{cases}$$

$$(8.9.12)$$

where α and σ are control functions, defined by the N-dimensional vectors $\{\sigma_i\}1 \le i \le N$, $\{\alpha_{1,i}\}1 \le i \le N$ and $\{\alpha_{2,i}\}1 \le i \le N$. These parameters α and σ appear in the pseudo-state equations exactly like λ and μ in the linearized state equations. The cost function to be minimized in order to obtain the sentinel attached to α_n or σ_n is

$$J(\sigma,\alpha) = \frac{1}{2}|C\rho|_H^2 - (\sigma_n \text{ or } \alpha_n)$$

$$(8.9.13)$$

and the corresponding sentinel is given by the "observation" of ρ at optimality:

$$w = C\rho.$$

$$(8.9.14)$$

The problem now is to calculate the two sentinels of the linearized problem (8.2.1). We apply the formulas of section 2 of chapter 1, concerning the method we qualified as direct. For a given point a:

1. We solve (8.2.1), respectively when $\hat{a} = e^1 = \begin{bmatrix} 1 \\ 0 \end{bmatrix}$ and $\hat{a} = e^2 = \begin{bmatrix} 0 \\ 1 \end{bmatrix}$, thus obtaining the solutions $\psi_1(x,t;a)$ and $\psi_2(x,t;a)$ and the corresponding observations of the elementary solutions:

$$C\psi_i(t;a) = \psi_i(x_k,t;a), \ i \in \{1, 2\}, \ k \in \{1, ...,M\}.$$

2. We calculate the components Λ_i^j of the 2×2 matrix Λ, for $1 \le i, j \le 2$:

$$\Lambda_i^j = \sum_{k=1}^{k=M} \int_0^T \psi_i(x_k,t)\psi_j(x_k,t)dt = \left(C\psi_i, C\psi_j \right)_H$$

3. We define γ^n by $\Lambda\gamma^n = e^n$, $n \in \{1, 2\}$ and $w^n(a)$ by $w^n(a) = B'(a)\gamma^n = \gamma_1^n\psi_1(x_k,t;a) + \gamma_2^n\psi_2(x_k,t;a)$

8.3.2 Indirect method

We solve the following optimal control problem: find the control $\alpha \in R^2$ that minimizes the cost function $J(\alpha)$ defined via the state ρ by

$$\rho' + A\rho = s(t)\ \delta_a'\alpha\ ,\ \rho(x, 0)=0;$$

control $\alpha = \{\alpha^j\}$, j from 1 to 2 ;

$$\text{cost } J(\alpha) = \frac{1}{2} \sum_{k=1}^{k=M} \int_0^T \rho^2(x_k, t;\alpha)\ dt - \alpha_{n)}$$

where the n-th sentinel is given by

$$w_k^n(t; a) = \rho(x_k, t), \ 1 \le k \le M. \tag{8.3.4}$$

Remark 8.3.1

The observation z is a (nonlinear) function of a: $z = B(a)$. Its differential at point a B' (a) transforms a change of position $\hat{a}$ into a change of observation $\hat{z}$

$$\hat{z} = B'(a)\hat{a} \tag{8.3.5}$$

Inversely, noting $W(a)=[w_1(a) ; w_2(a)]$, we have, from (8.3.1),

$$W(a)\hat{z} = \hat{a} , \tag{8.3.6}$$

so that the operator $W(a)$ transforms a change of observation $\hat{z}$ into a change of position $\hat{a}$ From (8.3.5) and (8.3.6) it results that, starting from a given $\hat{a}$, B'(a) transforms it in $\hat{z}$. Then, applying $W(a)$ to this element of H = $L^2(0, T; R^2)$ of the form $\hat{z} = B'(a)\hat{a}$, we come back to $\hat{a}$. It results that

$$W(a) \, B'(a) = Id_{R^2} = [e^1 \ e^2] \text{ the identity matrix of } R^2. \tag{8.3.7}$$

In the left hand side of (8.3.7) the term in i-th line and j-th column is the inner product, in the space of observations H, of w^i and $\dfrac{\partial B}{\partial a_j}$.

$$\left(w^i, \frac{\partial B}{\partial a_j} \right)_H = \sum_{k=1}^{M} \int_{]0,T[} w_k^i(t) \frac{\partial z_k}{\partial a_j}(t, a)dt.$$

where $z_k(t, a)$ is the exactly observed pollutant concentration at the k-th point of measurement, at time t and for the position a of the source.

9 Recapitulation

The aim of this chapter is to recapitulate the mathematics we encountered while discussing the method of sentinels. We successively recall important definitions, discuss inverse problems and their numerical solution, and end with a convergence result. Indeed, the mathematical requirements for our studies were at the level of matrix algebra in finite-dimensional spaces. The reason for that is we dealt with finite-dimensional approximations of the missing datas. This is justified by a result of convergence when these finite-dimensionnal approximations tend towards the true missing data. Anyway most of the time we use finite-dimensional approximations of the state equations, since we are using a finite element model of the state.

9.1 Definitions

Space V of parameters

This is the vector space R^N. Each vector v of V has N real components $v_1, \cdots, v_N$. V is equipped with the usual Euclidean inner product, denoted $(.,.)_V$:

$$(u,v)_V = \sum_{i=1}^{i=N} u_i v_i$$ (9.1.1)

Space H of observations

This is the Hilbert space $L^2(]0,T[;R^M)$ or $\left(L^2(0,T)\right)^M$ of M-tuples of square-integrable functions. Each element of H consists of M square integrable functions of time $z_1, \cdots, z_M$. Each of these M components z_k:]0, T[→R is a real-valued function of time t defined on the time interval]0, T[. This space H is equipped with the inner product

$$(\varphi, \psi)_H = \sum_{k=1}^{k=M} \int_0^T \varphi_k(t) \psi_k(t) dt$$ (9.1.2)

Operator B

This is a linear operator from V to H. This operator B is supposed to be one - to - one, i.e., equivalently,

$$\forall v, Bv = 0 \Rightarrow v = 0 \text{ or } \forall u \text{ and } v, Bu = Bv \Rightarrow u = v$$ (9.1.3)

This is the fundamental requirement for a sentinel to be defined. It is also a compulsory assumption for the system to be identifiable.

Proposition 9.1.1

The operator B is one-to-one if and only if the N measurements Be^i are linearly independent.

Proof $\sum_{i=1}^{i=N} v_i Be^i = B \sum_{i=1}^{i=N} v_i e^i = Bv = 0 \Leftrightarrow v=0.$

Linear system

This is the triple $S = (V, H, B)$.

Identifiability

We say that a system S is identifiable if the operator B is one-to-one. If an element z of H is the image by B of an element v of V, we say that the observation is exact. In the case where B is one-to-one this element v is the unique original of z for the application B. So we can speak of identifiability of v given z in range (B). We can view the vector v as a vector of input parameters and z as a vector of M output measurements in response to this input. Then we can say: zero output $\Rightarrow$ zero input.

Adjoint of B

This is the linear operator from H to V, denoted B^* and defined by

$$\forall z \in H \text{ and } \forall v \in V, \ \left(B^* z, v\right)_V = \left(Bv, z\right)_H \qquad (9.1.4)$$

Canonical basis of V

This is the family of N vectors $\{e^1, ..., e^N\}$, where e^i denotes the vector of V whose all components are zero, except the i-th one, which is 1:

$$e^i_j = \delta^i_j = \begin{cases} 1 \text{ if } j = i \\ 0 \text{ if } j \neq i \end{cases} \qquad (9.1.5)$$

Operator Λ

This is the operator B^*B. Since $B: V \rightarrow H$ and $B^*: H \rightarrow V$ then $B^*B: V \rightarrow V$. Consequently the operator Λ is represented by an $N \times N$ matrix Λ. It is well known that the i-th row and j-th column component of a matrix Λ is given by the inner product

$$\Lambda^j_i = \left(\Lambda e^i, e^j\right)_V = \left(Be^i, Be^j\right)_H. \qquad (9.1.6)$$

This matrix Λ is symmetric, as shown, for example, by (9.1.6). If B is one-to-one, the matrix Λ is positive definite, i.e.

Cost function J

In general, due to noise in measurements, the observation z is not exact. There is no v in V such that Bv = z. However, we can associate, with z obtained by measurements, a vector $\tilde{v}$ of V by the so-called least squares method, provided we can minimize a cost function

$$J(v) = \frac{1}{2} |Bv - z|_H^2 \qquad (9.1.7)$$

J(v) can be rewritten as

$$J(v) = \frac{1}{2}(Bv - z, Bv - z)_H = \frac{1}{2}(\Lambda v, v)_V - (B^* z, v)_V + \text{constant} \qquad (9.1.8)$$

Provided B is one-to-one, the matrix Λ is positive definite. It is well-known that under this condition on Λ

1 its inverse Λ^{-1} exists.

2 $\tilde{v} = \Lambda^{-1} B^* z$, minimizes J, i.e., $\forall v \in V, \ J(\tilde{v}) \le J(v)$.

9.2 Inverse problems

The direct problem consists of calculating the observation z = z(v) corresponding to the vector of parameters v:

Cause v → Effect z = Bv

The inverse problem

The inverse problem consists of calculating the components of the vector v of parameters from the observation z, that is, calculate a vector v of parameters that lead to an observation z:

Effect z → Cause v

A very simple result in this respect is the following. Suppose that we are given an exact observation z, that is to say, there is a vector v in V such that z = Bv. We wish to recover the i-th component v_i of this vector v by using an element w^i of H, more precisely an element in the range of B. Let

$$w^i = Bu^i \qquad (9.2.1)$$

denote this element. For the time being, u^i is at our disposal and can be taken anywhere in V. Our claim is that we can choose u^i so that for every exact observation z the inner product of w^i and z is equal

$$v_i = \left(w^i, z \right)_H \qquad (9.2.2c)$$

Proposition 9.2.1

Let us successively define u^i and w^i by (9.2.2a) and (9.2.2b). Then for every exact observation z the value of the i-th component of the original vector v is given by (9.2.2c).

$$v_i = \left(w^i, z \right)_H \qquad (9.2.3)$$

Definition 9.2.1

w^i is called the sentinel associated with the parameter v_i.

Remark 2.1

Once the sentinel w^i has been calculated by (9.2.2a) and (9.2.2b), formula (9.2.3) enables us to monitor the parameter v^i. Hence the name of *sentinel* for w^i, from Italian *sentinella*, from Latin *sentire* (to behave like a sensor).

Remark 2.2

The vector v does not depend upon the observation z. Therefore we are tempted to still use formulas (9.2.2a), (9.2.2b) and (9.2.2c) to get a solution of the inverse problem when the observation is not exact. Indeed, we are now going to show that a least squares solution of the inverse problem results in the same formulas.

Let z be an element of H, not necessarily in the range of B, and J be the cost function

$$J(v) = \frac{1}{2} | Bv - z |_H^2 \qquad (9.2.4)$$

We have:

Proposition 9.2.2

There is one and only one element v in V minimizing J. Its components v_i are given by (9.2.3), where w^i is defined by (9.2.2a) and (9.2.2b).

Proof:

We know (see section 9.1) that J is minimized by $u = \Lambda^{-1}B^*z$. Its i-th component is given by $u_i = (e^i, u)_V = (e^i, \Lambda^{-1}B^*z)_V = (\Lambda^{-1}e^i, B^*z)_V = (u^i, B^*z)_V = (Bu^i, z)_H = (w^i, z)_H$.

Remark 9.2.3

The difference between propositions (9.2.1) and (9.2.2) is that the former needs z to be given in BV, i.e. the range of B, i.e., the image of V by B, whereas in the latter z can be given anywhere in H, not necessarily in range(B). In both cases we associate with an element z in H a vector v in V by equation (9.2.3).

Remark 9.2.4

Proposition 9.2.2 enables us to introduce the so-called pseudo inverse operator W of B. It is this continuous linear operator that transforms each element $z \in H$ into that element u whose components are given, for i = 1 to N, by $u_i = \left(w^i, z\right)_H = (Wz)_i$, where $Wz = \left(w^1z, \cdots, w^Nz\right)$. We say that u is the pseudo inverse of z. We identify the pseudo inverse operator W and $\left(w^1, \cdots, w^N\right)$: $W = \left(w^1, \cdots, w^N\right)$.

9.3 A convergence result

Example 9.3.1 Time history of the flow rate

In the case of section 4 of chapter 1 for example the flow rate $s \in L^2(0,T)$ is the pollution term and the initial concentration of pollutant $y^0 \in L^2(\Omega)$ is the missing term. Let $V = L^2(0,T) \times L^2(\Omega)$ denote the space of inputs $v = (\lambda, \tau)$, and B: V→H the operator that responds to an input v by an output $z \in H = L^2(]0,T[\times R^M)$ via the solution of the system

$$
\begin{cases}
y' + Ay = s(t)\delta(x-a) & \text{on } Q = \Omega \times]0,T[\\
y(x,0) = y^0(x) & \text{on } \Omega \\
\dfrac{\partial y}{\partial n} = 0 & \text{on } \Sigma = \Gamma \times]0,T[
\end{cases}
$$

(9.3.1.)

$$z_k(t) \equiv y\left(x_k, t\right)$$

In this case V is infinite dimensional and K = BV, the image of V by B, has no reason to be closed in H. However in V the semi-norm, denoted $|.|_F$, and defined by

$$|\lambda|_F = |B\lambda|_H \qquad (9.3.2)$$

actually is a norm because $|\lambda|_F = 0 \Rightarrow |B\lambda|_H = 0 \Rightarrow B\lambda = 0 \Rightarrow \lambda = 0$, thanks to the injectivity of B. Let F be the completion of V for that norm, and still denote by B the extension to F (equipped with the norm $|.|_F$) of the isometry from V into H. Then BF, complete subspace of H, is closed in H and we can speak of the projection of H on BF (as illustrated by figure 1.3 of chapter 1). We define the operator W : H→F by $\lambda = Wz \Leftrightarrow B\lambda$ is the projection of z on BF $\Leftrightarrow \lambda$ is that element of V such that the distance from z to $B\lambda$ is minimum $\Leftrightarrow \lambda$ minimizes

$$J(\lambda) = \frac{1}{2}|B\lambda - z|_H^2 \qquad (9.3.3)$$

Let $b_1, ..., b_2, ... , b_m, ...$ denote a basis of F, in the following sense:

The linear and finite combinations $\underset{finite}{\sum} \tau_j b_j \; \tau_j \in R$ are dense in F $\qquad$ (9.3.4)

For example a basis of $L^2(0,T)$ consists of the functions $\left\{ \sin n\frac{2\pi}{T}t \right\}_{n\geq1}$.

Let $F_m = [b_1, ..., b_2, ... , b_m]$ denote the subspace of F spanned by $b_1, ..., b_m$.

Let u and u_m, respectively, denote the solutions of the problems

$$\underset{v\in F}{\inf} J(v) \qquad (9.3.5)$$

$$\underset{v\in F_m}{\inf} J(v) \qquad (9.3.6)$$

Proposition 9.3.1

With the above notations and hypotheses we have the following convergences:

$$u_m \to u \text{ in } F \text{ as } m \to \infty \qquad (9.3.7)$$

$$Bu_m \to Bu \text{ in H strongly as } m \to \infty \qquad (9.3.8)$$

Proof

We have the optimality conditions

$$(J'(u), v) = 0 \ \forall v \in F$$

and

$$(J'(u_m), v_m) = 0, \ \forall \ v_m \in F_m$$

which can be rewriten as

$$(Bu, Bv)_H - (Bv, z)_H = 0 \ \forall v \in V \qquad (9.3.9)$$

and

$$(Bu_m, Bv)_H - (Bv, z)_H = 0 \ \forall v \in V_m \qquad (9.3.10).$$

Taking $v = u_m$ in (9.3.10) we obtain

$$\left| Bu_m \right|^2_H = (Bu_m, z)_H \leq \left| Bu_m \right|_H \left| z \right|_H, \text{ whence } \left| Bu_m \right|_H \leq \left| z \right|_H.$$

Therefore we can extract a subsequence, still denoted Bu_m, and such that

$$Bu_m \to q \text{ in H weakly} \qquad (9.3.11)$$

For each b_i, we have

$$\text{for } m \geq i, (Bu_m, Bb_i)_H - (Bb_i, z)_H = 0$$

Therefore by passing to the limit we have

$$(q, Bb_i)_H - (Bb_i, z)_H = 0$$

Since we also have $(Bu, Bb_i)-(Bb_i, z)=0$, we have $(q, Bb_i)_H = (Bu, Bb_i)_H$

But B being injective (it even is an isometry!), the $\{Bb_i\}$ constitute a basis of BF, from which it results that

$$q = Bu \tag{9.3.12}$$

From (9.3.11) and (9.3.12) we have

$$Bu_m \to Bu \text{ in } H \text{ weakly} \tag{9.3.13}$$

We even have

$$Bu_m \to Bu \text{ in } H \text{ strongly.} \tag{9.3.14}$$

Indeed, we have

$$(Bu_m - Bu, Bu_m - Bu)_H = (Bu_m, Bu_m)_H - (Bu_m, Bu)_H - (Bu, Bu_m - Bu)_H$$

$$(Bu_m, Bu_m)_H = (Bu_m, z)_H \to (Bu, z)_H \quad \text{from (9.3.13).}$$

$$-(Bu_m, Bu)_H \to -(Bu, Bu)_H = -(Bu, z)_H \quad \text{from (9.3.13) and (3.9).}$$

$$(Bu, Bu_m - Bu)_H \to 0 \quad \text{from (3.13).}$$

whence (3.14). As an immediate consequence and since B is an isometry from F to BF,

$$u_m \to u \text{ in } F \text{ strong} \tag{9.3.15}$$

9.4 Gauss-Newton method

A so-called Gauss-Newton method well adapted to our problem is the following:

$$\begin{cases} v^0 \text{ initial guess} \\ p^n = -W^n(B^n - z_d) \\ v^{n+1} - v^n = p^n \end{cases} \tag{9.4.1}$$

where $B^n = B(v^n)$ and W^n is the generalized inverse of the linearized operator $B'\left(v^n\right)$:

$$W^n = \left(\Lambda^n\right)^{-1} B'^{n*} \tag{9.4.2}$$

where $\Lambda^n = B'^{n*} B'^n$

The gradient of J is

$$J'(v) = B'(v)^* (B(v) - z_d).$$ (9.4.3)

If u minimizes J,

$$\begin{cases} u \in R^N \\ B'(u)^* (B(u) - z_d) = 0 \end{cases}$$ (9.4.4)

The linearized operator of B is $B'^n = B'(v^n)$.

1 $J(v^{n+1}) < J(v^n)$ **and** $J(v^n) \to \mu \geq 0$

$$pn = -Wn(Bn - zd) = -\left(\Lambda^n\right)^{-1} B'^{n*} (Bn - zd) = \left(\Lambda^n\right)^{-1} (-gn)$$

where $g^n = J'(v^n)$

$$p^n = \left(\Lambda^n\right)^{-1} \left(-g^n\right)$$ (9.4.5)

$$-g^n = \Lambda^n p^n$$ (9.4.6)

The matrix $\left(\Lambda^n\right)^{-1}$ being symmetric and positive definite, from equality (9.4.6) it results that

$$((-g^n), p^n)_V = (-g^n, \left(\Lambda^n\right)^{-1}(-g^n)) > 0,$$

unless $g^n = 0$, in which case $v^n = u$, u being a solution to (9.4.4).

Therefore $p^n = v^{n+1} - v^n$ is a descent direction, that is to say that the iterates $\{v^n\}$ converge downhhill to a minimum of J (i.e., to a solution of $\nabla J = 0$)

$$J(v^{n+1}) < J(v^n).$$ (9.4.7)

The sequence $\{J(v^n)\}$, being decreasing and bounded below by 0 admits a limit $\mu \geq 0$.

$$J(v^n) \to \mu \geq 0.$$ (9.4.8)

2 $B(v^n) \to B(u)$, $u \in V$

Let us use the information we have. Let

$$z^n = B(v^n).$$

(9.4.9)

(9.4.8) and (9.4.9) imply that $J(v^n) \leq J(v^0)$ and therefore z^n is bounded in H.

Therefore we can extract a subsequence, still denoted $\{z^n\}$, that converges weakly to $z \in H$. Then this implies that $|z - z_d| \leq \lim \inf |z^n - z_d|_H$ $= \mu$. .If $|z - z_d|_H < \mu$ then for n large enough $J(v^n) < \mu$. But that is impossible. Therefore we have $|z - z_d|_H = \mu$.

Next we know that if

$$\begin{cases} \|z^n - z_d\| \to z - z_d \text{ weakly} \\ \|z^n - z_d\| \to \|z - z_d\| \end{cases} \Rightarrow z^n - z_d \to z - z_d \text{ strongly}$$

then z_n tends strongly to z.

This does not necessarily imply z of the form $z = Bu$. But if we introduce the hypothesis *the image of V by B is closed,* then we have the following consequence: If $B(v^n) \to z$, z must be of the form $z = B(u)$, $u \in V$.

The iterates (9.4.1) always define downhill paths that, hopefully, converge to a vector u for which we have (9.4.4). The question of convergence will not be pursued here except to remark that it results from the construction of u, that it is that element of V for which the distance in H between z_d and the manifold $z = B(v)$ is minimum. That is characterized by the orthogonality of $B(u) - z_d$ and the vectors $B'(u)\ \hat{v}$, $\hat{v} \in V$ which span the tangent plane to the manyfold. This orthogonality is equivalent to $J'(u) = 0$:

$$B(u) - z_d \perp B'(u)\ \hat{v}, \forall\ \hat{v} \in V \Leftrightarrow (B(u) - z_d, B'(u)\ \hat{v})_H = 0\ \forall\ \hat{v} \in V$$

$$\Leftrightarrow (B'^*(u)(B(u) - z_d), \hat{v})_V = 0\ \forall\ \hat{v} \in V \Leftrightarrow B'^*(u)(B(u) - z_d) = 0. \Leftrightarrow$$
$$J'(u) = 0.$$

10 Shallow waters

We present a modelling of pollution in shallow waters (bays, estuaries, gulfs, lakes, and rivers), taking into account tide propagation and the dispersion of pollutants. Then we apply the sentinels method to this model. For the model and its finite elements approximation we refer to C.Taylor, J.M. Davis[1], and A. Bermudez and M.E. Vaquez [1].

10.1 The movement of tides; Saint-Venant shallow water equations

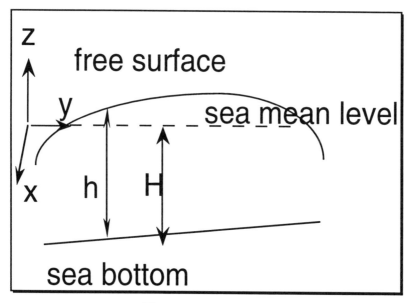

Figure 10.1

For the sake of completeness we present below the Saint-Venant shallow water equations.

$$\frac{\partial U}{\partial t} + \vec{c} \bullet \vec{\nabla} U - \omega V + g\frac{\partial(h - h')}{\partial x} + g\frac{\left(U^2 + V^2\right)^{\frac{1}{2}}}{C^2 H_1} - \frac{\tau_x}{H_1} = 0$$

(10.1.1)

$$\frac{\partial V}{\partial t} + \vec{c} \bullet \vec{\nabla} V + \omega U + g \frac{\partial(h - h')}{\partial y} + g \frac{\left(U^2 + V^2\right)^{\frac{1}{2}}}{C^2 H_1} - \frac{\tau_y}{H_1} = 0 \qquad (10.1.2)$$

$$\frac{\partial h}{\partial t} + \vec{\nabla} \bullet \vec{q} = 0$$

here H_1 = h, C = Chézy friction coefficient, and $\vec{c}$ = (U,V) . $\qquad (10.1.3)$

$$q_i = hu_i, i = 1, 2, u_1 = U, \ u_2 = V$$

The boundary conditions are

h = h_1 (t) on Γ_h (open sea specified water heigth) $\qquad (10.1.4)$

$\hat{q} \bullet \hat{n} = q_n^* \neq 0$ on Γ_0 (incoming river water) $\qquad (10.1.5)$

$\vec{q} \bullet \vec{n}$ = 0 on Γ_2(totally reflecting boundaries) $\qquad (10.1.6)$

Let us change the notations and write these equations under the form

$$\frac{\partial w}{\partial t} + \frac{\partial F_1}{\partial x_1} + \frac{\partial F_2}{\partial x_2} = 0 \qquad (10.1.7)$$

where $\quad w = \begin{vmatrix} h \\ q_1 \\ q_2 \end{vmatrix},$ $\qquad (10.1.8)$

$$F1 = \begin{bmatrix} q1 \\ \dfrac{q_1^2}{h} + \dfrac{1}{2}gh^2 \\ \dfrac{q_1 q_2}{h} \end{bmatrix},$$
(10.1.9)

$$F2 = \begin{bmatrix} q2 \\ \dfrac{q_1 q_2}{h} \\ \dfrac{q_2^2}{h} + \dfrac{1}{2}gh^2 \end{bmatrix},$$
(10.1.10)

$$G = \begin{bmatrix} \omega q_2 - g\dfrac{\left(q_1^2 + q_2^2\right)^{\frac{1}{2}}}{hC^2} - gh\dfrac{\partial H}{\partial x_1} + \tau_x \\ -\omega q_1 - g\dfrac{\left(q_1^2 + q_2^2\right)^{\frac{1}{2}}}{hC^2} - gh\dfrac{\partial H}{\partial x_2} + \tau_y \end{bmatrix}$$
(10.1.11)

$$\frac{\partial h}{\partial t} + \nabla q = 0$$
(10 1.12)

$$\frac{\partial q_1}{\partial t} + \frac{\partial}{\partial x_1}\left(\frac{q_1^2}{h} + \frac{1}{2}gh^2\right) + \frac{\partial}{\partial x_2}\left(\frac{q_1 q_2}{h}\right) - \omega q_2 +$$
$$g\frac{\left(q_1^2 + q_2^2\right)^{\frac{1}{2}}}{C^2 H_1} = gh\frac{\partial H}{\partial x_1} + \tau_x$$
(10.1.13)

$$\frac{\partial q_2}{\partial t} + \frac{\partial}{\partial x_2}\left(\frac{q_2^2}{h} + \frac{1}{2}gh^2\right) + \frac{\partial}{\partial x_1}\left(\frac{q_1 q_2}{h}\right) + \omega q_1 +$$
$$g\frac{\left(q_1^2 + q_2^2\right)^{\frac{1}{2}}}{C^2 H_1} = gh\frac{\partial H}{\partial x_2} + \tau_y$$
(10.1.14)

10.2 Numerical solution of shallow water equations

We solve by the modified Euler method the equations resulting from a discretization in space by the finite elements method (H.Ninomiya and K.Onishi[1])

10.3 Weak formulation of the problem

For the exact problem we multiply (10.1.12) by a test function $\hat{h}$

$$\int_\Omega \hat{h}\left(\frac{\partial h}{\partial t} + \vec{\nabla} \bullet \vec{q}\right) d\Omega = 0$$

and we integrate by parts. But

$$\int_\Gamma \hat{h}\,\vec{q} \bullet \vec{n}\, d\Gamma = \int_{\Gamma_h} \hat{h}\ \text{q.n}\ d\Gamma + \int_{\Gamma_2} \hat{h}\ \text{q.n}\ d\Gamma + \int_{\Gamma_0} \hat{h}\ \text{q.n}\ d\Gamma = \int_{\Gamma_0} \hat{h}\ q_n^*\ d\Gamma$$

The first integral is null because $\hat{h}$ is null on Γ_h, the second one because $\vec{q} \bullet \vec{n} = 0$ on Γ_2. Finally $\vec{q} \bullet \vec{n}$ is given on Γ_0

Therefore the exact problem can be formulated as shown below

One looks for h given on Γ_h and satisfying, for every test function $\hat{h}$ null on Γ_h,

Dispersion, convection, and reaction equations

$$\int_\Omega \hat{h}\frac{\partial h}{\partial t}\,d\Omega - \int_\Omega \vec{\nabla}\hat{h} \bullet \vec{q}d\Omega + \int_{\Gamma_0} \hat{h}q_n^*d\Gamma = 0.$$

10.4 Reaction-convection-dispersion equations

For the pollution due to chemical species, we have the reaction-convection-dispersion equations. The concentration of a species, say c_i, depends upon convection and dispersion of c_i, and reactions of c_i with c_j, $j \neq i$.

Assuming a "well-mixed" estuary, and the validity of integrating with respect to z, we can define the depth-averaged concentration

$$h = \frac{1}{H_1} \int_{-H}^{h} c_i \, dz. \tag{10.4.1}$$

We have

$$\frac{\partial y}{\partial t} + \bar{u} \bullet \vec{\nabla} y - \frac{1}{h} \nabla(Dh\nabla y) + \sigma y = \sum_{i=1}^{i=N_1} \lambda_i s_i(t)\delta(x - a_i) \tag{10.4.2}$$

where

$$\bullet \, \nabla(Dhy) = \sum_{i=1}^{i=2} \frac{\partial}{\partial x_i} \left(D_i H_i \frac{\partial y}{\partial x_i} \right)$$

$$\bullet \, D = \begin{bmatrix} D_1 & 0 \\ 0 & D_2 \end{bmatrix}$$

$\bullet D_1$ and D_2 are coefficients defined by

$$D_i = \frac{1}{H_1} \int_{-H}^{h} e_i \, dz \qquad i=1, 2$$

where e_i is the eddy diffusion coefficient in direction i.
The boundary conditions are

$$y = 0 \text{ on } \Gamma_1 \text{ (open sea)} \tag{10.4.3}$$

$$\nabla y_n = y_n^* \neq 0 \text{ on } \Gamma_2 \text{ (incoming river water)} \tag{10.4.4}$$

$$n \bullet Dh\nabla y = \sum_{i=1}^{i=2} n_i \left(D_i h \frac{\partial y}{\partial x_i} \right) = 0$$

on Γ_3 (totally reflecting boundaries) \qquad (10.4.5)

$$\nabla y.n = y_n^* \neq 0 \text{ on } \Gamma_5 \text{ (outgoing river water)} \tag{10.4.6}$$

The initial conditions are

$$y(x,0) = 0 \text{ for } x \in \Omega \tag{10.4.7}$$

Remark 4.1

More generally we could suppose that

$$y(x,0) = \sum_{j=1}^{j=N_2} \tau_j \chi_j(x) \tag{10.4.8}$$

where $\chi_j(x)$ is the characteristic function of the j-th element of a finite element mesh or of the j-th superelement, each superelement regrouping several elements. In effect in applications the number of elements may be large and it may be convenient to reunify several elements into a singlesuper element. Let N_2 be the number of elements or super elements. According to whether we model the initial condition by (10.4.7) or (10.4.8), the matrix Λ is $N \times N$ with $N = N_1$ or $N = N_1 + N_2$. Of course the latter choice is better than the former, even with N_2 small we design sentinels practically insensitive to initial conditions, whereas with $N_2 = 0$ there is no reason for the obtained sentinels to be insensitive to initial conditions (3.6.1).

10.5 Sentinels

Defining the operator A by

$$Ay = \vec{u} \bullet \vec{\nabla} y - \frac{1}{h} \nabla (Dh\nabla y) + \sigma y \tag{10.5.1}$$

we are in a situation similar to that already encountered in Chapters 1, 2, 3, and 4:

$$y' + Ay = \sum_{i=1}^{i=N_1} \lambda_i s_i(t) \delta(x - a_i) \tag{10.5.2}$$

$$y(0) = 0.$$

10.5.1 The state of the system

It is given by (10.2.1) - (10.2.6) for water height h and velocity (U,V). It is given by (10.5.2) and (10.4.3) - (10.4.7) for pollutant concentration y. First we determine the fields h, U and V, once for all, either by computations or by measurements. It is not our purpose here to enter into the technical details of using a finite difference or a finite element scheme to determine them, for which we refer to the bibliography. Then we determine for $i \in [1, ..., N]$ the solution $y = \Psi_i$ of

$$y' + Ay = s_i(t)\, \delta(x-a_i)$$

$$y(0) = 0.$$

The state of the system is given by

$$y = \sum_{i=1}^{i=N} \lambda_i \Psi_i \tag{10.5.3}$$

10.5.2 State Observation

We have at our disposal our favorite measurement system, which consists in M sensors placed at points x_k, $k \in [1, ..., M]$, each one providing a continuous signal $z_k(t)$ on the time interval [0,T]. These M measurements constitute the observation and lie in the space $(L^2(0,T))^M$.

10.5.3 Building a sentinel

For example our aim is to build a sentinel to identify the total quantity of pollutant deversed by the outfalls. We recall that, with our usual notations,

$$(w,z)_H = (w, Cy) = (C*w, y) = (-q' + A*q, y) = (q(0), y(0))_{L^2(\Omega)} + (q, y' + Ay)_{L^2(Q)}.$$

Assuming y(0) = 0, we have

$$\left(w, z\right) = \left(q, \sum_{i=1}^{i=N} \lambda_i s_i(t)\delta(x - a_i)\right) = \sum_{i=1}^{i=N} \lambda_i \int_0^T s_i(t)q(a_i, t)dt =$$

$$\sum_{i=1}^{i=N} \lambda_i \int_0^T s_i(t)dt$$

provided the function q satisfies

$$\int_0^T s_i(t) \, q(a_j,t) \, dt = \int_0^T s_i(t) \, dt, \ i=1,\dots,N_1. \tag{10.5.4}$$

This condition is fullfilled by the adjoint state of the following optimal control problem:

State defined by

$$\rho' + A\rho = \sum_{1=I}^{N} \alpha_i s_i(t) \delta(x - a_i) \tag{10.5.5}$$

Cost function

$$J(\alpha) = \frac{1}{2} \|C\rho\|_H^2 - (c,\alpha)_V \tag{10.5.6}$$

$$c_i = \int_0^T s_i(t) dt, \ 1 \le i \le N. \tag{10.5.7}$$

The adjoint state is defined by

$$-q' + A^* q = C^* w, \ q(T) = 0, \tag{10.5.8}$$

and the optimality conditions are

$$\int_0^T s_i(t) \, q(a_j,t) \, dt = \int_0^T s_i(t) \, dt \ .$$

A Appendix:

A 1 Sentinels with a given sensitivity and duality

(IN COLLABORATION WITH J.L.LIONS)

A 1 1 Introduction.

We return in this appendix to one of the situations studied in chapter
1 to bring two types of complements:
1. a duality formula
2. the introduction of functional spaces necessary for understanding the
problem when one has no information about the initial data
We choose an example, but what follows is completely general and
adapts to all the situations studied in this book.

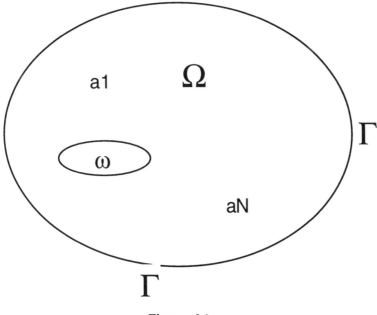

Figure A1

We consider the open set $\Omega \subset R^n$, n = 2 or 3 (the case n = 1 presents
additional technical difficulties, artificially created by the lack of "physical
realism " of this case. Anyway, it is only a question of technical difficulties
that, at the expense of additional developments, does not affect what is
essential.

The state equation is given by

$$y' + Ay = \sum_{i=1}^{i=n} \lambda_i s_i(t) \delta\left(x - a_i\right) \tag{A.1.1}$$

where

$$\begin{cases} A \text{ is a second order elliptic operator} \\ \{a_i\} \text{ is given} \\ \text{the } s_i(t) \text{ are given in } L^2(0,T) \end{cases}$$

and where, on the other hand, the λ_i are not given: they are the numbers we are looking for.

The initial conditions are unknown.

We pose

$$y(0) = y^0 + \tau \, \hat{y}^0 \tag{A.1.2a}$$

with y^0 given in $L^2(\Omega)$, and $\hat{y}^0$ any element in the ball of $L^2(\Omega)$ of center 0 and radius 1.

Remark A.1.1

In chapter 1, one considers the case where $y^0 = 0$ and where $\hat{y}^0$ is replaced by

$$\sum_{j=1}^{j=N_2} \tau_j \hat{y}_j^0 \tag{A.1.2b}$$

the $\hat{y}_j^0$ being given in $L^2(\Omega)$. To say it differently, we have informations about the structure of $\tau \hat{y}^0$, which is not the case for (A.1.2a).

One adds to (A.1.1), (A.1.2a) a boundary condition. To fix ideas, one takes, as in Chapter 1,

$$\frac{\partial y}{\partial v_A} = 0,$$

$$\frac{\partial y}{\partial v_A} \text{ being the conormal derivative associated with A.}$$

Remark A.1.2

What follows adapts also to the case where y is known on $\omega \times \,]0,T[$ with addition of a noise. As a matter of fact, the sentinels we are going to define and calculate only depend upon the position of the points a_i and of the observatory ω and not at all on the values observed on ω.

Now we are given an observatory as in Figure A1, ω being an open set arbitrarily "small" , contained in Ω , and containing no source point a_i.

We suppose that y is measured on $\omega \times \,]0,T[$.We therefore know

$$y \text{ on } \omega \times \,]0,T[\tag{A.1.4}$$

We now look for a sentinel under the following form

$$S\left(\lambda, \tau, \, \hat{y}^0\right) = \int_{\omega \times]0,T[} w(x,t)y(x,t;\lambda, \tau, \, \hat{y}^0)dxdt \tag{A.1.5}$$

where w is to be determined in $L^2(\omega \times \,]0,T[)$ and $y(x,t;\lambda, \tau, \, \hat{y}^0)$ denotes the solution of (A.1.1), (A.1.2), and (A.1.3) corresponding to given values $\lambda = \{\lambda_i\}_{1 \le i \le N}$ and to a known perturbation $\tau\hat{y}^0$.

One tries to determine - if possible - w(x, t) such that

$$\begin{cases} \dfrac{\partial S}{\partial \tau}(0,0) = 0, \forall \hat{y}^0 \in L^2(\Omega) \\[2ex] \dfrac{\partial S}{\partial \lambda_i}(0,0) = \delta_n^i, \text{n given}, 1 \le n \le N, 1 \le i \le N \end{cases} \tag{A.1.6}$$

where (0, 0) stands for $\lambda = 0$ and $\tau = 0$

Remark A.1.3

If we can find w with (A.1.6), we then say that the sentinel (A.1.5) has a given sensibility with respect to the parameter λ_n . This sensibility is to determine. Cf. J.L.Lions [1]. The first condition in (A.1.6) expresses the insensibility of the functional with respect to the unknown initial condition. The second condition in (A.1.6) expresses the insensibility of the functional with respect to the unknown intensity of the i-th source of pollution if $i \ne n$, and the sensibility with respect to the n-th parameter λ_n, unknown too, which we want to determine.

Remark A.1.4

If there exists a function w such that (A.1.6) holds, then, in fact, there exists an infinity of such functions, which we denote K and we make the problem more precise by looking for

$$\inf_{K} \frac{1}{2} \int_{\omega \times]0,T[} w^2 dxdt \qquad (A.1.7)$$

Remark A.1.5

In the first introduction sentinels were presented in a slightly different way. We start from a function h_0 given in $L^2(\omega \times]0,T[)$ and we consider the functional (cf. (A.1.5))

$$S(\lambda, \tau\hat{y}^0) = \int_{\omega \times]0,T[} (h_0(x,t) + w(x,t))y(x,t;\lambda,\tau\hat{y}^0)dxdt \qquad (A.1.8)$$

where w is to be determined such that (A.1.6) holds.

A 2 Duality

We first transform this problem, as done in Chapter 2. We introduce to this effect the function q solution to

$$-q'+A^* q = w \chi_\omega \text{ in } \Omega \times]0,T[\qquad (A.2.1)$$

where A^* is the adjoint of A and where χ_ω denotes the characteristic function of ω. One adds to (2.1) the conditions

$$\begin{cases} q(x,T) = 0 \\ \dfrac{\partial q}{\partial v_{A^*}} = 0 \end{cases} \qquad (A.2.2)$$

($\dfrac{\partial q}{\partial v_{A^*}}$ being the conormal derivative associated with A^*). We then immediately check (as in Chapter 2) that the conditions (A.1.6) are, respectively, equivalent to

$$q(x, 0) = 0 \text{ (a.e.) in } \Omega \qquad (A.2.3)$$

$$\int_0^T s_i(t) q(a_i,t) dt = \delta_{i_0}^i, \ i_0 \text{ given}, \ 1 \le i \le N. \qquad (A.2.4)$$

We therefore look for w solution of (A.2.1) and (A.2.2) with (A.2.3) and (A.2.4).

We can formulate this problem in the following way:

For w chosen in $L^2(\omega \times]0,T[$), the problem (A.2.1) and (A.2.2) admits a unique solution q(x, t; w). We then define a linear operator L by

$$Lw = \{q(x, 0 ; w) ; \int_0^T q(a_i, t; w) \, s_i(t) \, dt, \, 1 \le i \le N.\} \qquad (A.2.5)$$

This operator is continuous linear from $L^2(\omega \times]0,T[ ) \to L^2(\Omega) \times R^N$

$$L \in \mathcal{L}(L^2(\omega \times]0,T[), L^2(\Omega) \times R^N) \qquad (A.2.6)$$

Indeed, for w given in $L^2(\omega \times] 0,T[)$, the solution of (A.2.1) and (A.2.2) verifies (at least if the coefficients of A and the boundary of Ω are regular enough)

$$q \in L^2(] 0,T[; H^2(\Omega)) \qquad (A.2.7)$$

and, according to the Sobolev imbedding theorem, we have $H^2(\Omega) \subset C^0(\overline{\Omega})$ {continuous functions in $\overline{\Omega}$} if the dimension of space is ≤ 3 (which is the case!). Therefore, $q(a_i, t) \in L^2(] 0,T[$ and one has (A.2.6).

We now introduce two convex functions

$$F_2(g,\alpha) = \begin{cases} 0 \text{ if } g = 0 \text{ and } \alpha = e_{i_0} = \{\delta_{i_0,i}\} \, 1 \le i \le N, i_0 \text{ given} \\ \infty \text{ else} \end{cases} \qquad (A.2.8)$$

F_2 being a proper convex function defined on $L^2(\Omega) \times R^N$ \qquad (A.2.9)

With these notations the problem (A.1.8) with (A.2.3) and (A.2.4) can be set forth as follows: We look for

$$\inf_{w \in L^2(\omega \times (0,T))} (F_1(w) + F_2(Lw)) \qquad (A.2.10)$$

This is just another way to express the problem, but it now enables us to apply the duality theory of W. Fenchel and T.R. Rockafellar [1] (cf. I. Ekeland and R. Temam [1]). We introduce the conjugate functions

$$F_1^*(w) = \sup_{\hat{w}} \left\{ \int_{\omega \times]0,T[} w\hat{w}\,dx\,dt - F_1(\hat{w}) \right\} = F_1(w)$$

$$F_2^*(g,\alpha) = \sup_{\hat{g},\hat{\alpha}} \left\{ \int_{\Omega} g\hat{g}\,dx + \alpha\hat{\alpha} - F_2(\hat{g},\hat{\alpha}) \right\} = \alpha_{i_0}$$

We then introduce the adjoint L^* of L:

$$L^* \in \mathbf{L}(L^2(\Omega) \times R^N ; L^2(\omega \times]0,T[), \tag{A.2.11}$$

which is given in the following way. For $(g, a) \in L^2(\Omega) \times R^N$ one defines φ by

$$\begin{cases} \varphi' + A\varphi = \sum_{i=1}^{i=N_1} \alpha_i s_i(t)\delta(x - a_i) \\ \varphi(x,0) = g(x) \\ \dfrac{\partial \varphi}{\partial \nu_A} = 0 \end{cases} \tag{A.2.12}$$

which gives a function $\varphi(x,t; g, \alpha)$. If we multiply (A.2.12) by $q = q(x, t; w)$, xe obtain, after integrations by parts

$$\sum_{i=1}^{i=N} \alpha_i \int_0^T s_i(t)q(a_i,t)\,dt = -\int_{\Omega} q(x,0)g(x)\,dx + \int_{\omega \times]0,T[} \varphi w\,dx\,dt$$

that is to say

$$\langle Lw|\{g,\alpha\}\rangle = \int_{\omega \times]0,T[} \varphi w\,dx\,dt \tag{A.2.13}$$

where the bracket in A.2.13) expresses the inner product in $L^2(\Omega) \times R^N$. It results from (2.13) that

$$L^*\{g, a\} = \varphi\chi_{\omega} \tag{A.2.14}$$

The duality theorem gives

$$\inf_{w \in L^2(\omega \times (0,T))} (F_1(w) + F_2(Lw)) =$$

$$- \inf_{g,\alpha} \left(F_1^*\left(L^*\{g,\alpha\}\right) + F_2^*(-g,-\alpha) \right) \tag{A.2.15}$$

that is to say, making it explicit:

$$\inf_{w \in L^2(\omega \times (0,T))} \left(\frac{1}{2} \int_{\omega \times]0,T[} w^2 \, dxdt \right) = -\inf_{g,\alpha} \left(\frac{1}{2} \int_{\Omega \times]0,T[} \varphi^2 \, dxdt - \alpha_{i_0} \right) \qquad \text{(A.2.16)}$$

where φ is the solution of (A.2.12), with (A.2.3) and (A.2.4).

Remark A2.1

In the case of Chapter 1, we have informations on the initial data $\hat{y}^0$, so that equation (2.12) for j is replaced by

$$\begin{cases} \rho' + A\rho = \displaystyle\sum_{i=1}^{i=N_1} \alpha_i s_i(t)\delta(x - a_i) \\[2mm] \rho(x,0) = \displaystyle\sum_{j=1}^{j=N_2} \beta_j \hat{y}_j^0 \\[2mm] \dfrac{\partial \varphi}{\partial v_A} = 0 \end{cases} \qquad \text{(A.2.17)}$$

Otherwise said, g then varies (in the notations of (A2.12)) in a finite - dimensional space.

Remark A.2.2

When g varies in a finite - dimensional space the existence of a solution to the dual problem

$$\inf \left\{ \frac{1}{2} \int_{\omega \times]0,T[} \varphi^2 \, dxdt - \alpha_{i_0} \right\} \qquad \text{(A.2.18)}$$

is immediate. It is much less simple if g varies in an infinite - dimensional space, see Section A.3)

Remark A.2.3

If φ_{opt} is the solution to (2.18), then the solution to

$$\inf \left\{ \frac{1}{2} \int_{\omega \times]0,T[} w^2 \, dxdt \right\} \text{ with (A.2.3) and (A.2.4) is given by}$$

$$w_{opt} = \varphi_{opt} \text{ on } \omega \times]0,T[\qquad \text{(A.2.19)}$$

194

Remark A 2.4

The employed numerical algorithms are relative to the direct solution to (2.18).

Remark A 2.5

The initial problem of the search of $\inf\left\{\dfrac{1}{2}\displaystyle\int_{\omega\times]0,T[}w^2\,dxdt\right\}$ with (A.2.3) and (A.2.4) can be approached (cf. J.L. LIONS [1]) by penalization, which leads to an optimality system that we find again, if we wish, by the optimality system obtained from the dual formulation, simpler because there are no more constraints on the state like in the primal formulation.

A.3 Existence of a solution and functional spaces

What follows is given as a complement and only arises in the case where we have at our disposal no observation about the "initial pollution". We can write (A.2.18), with the notations of Section 2, under the form

$$\inf\left\{\frac{1}{2}\left|L^*(g,\alpha)\right|^2_{L^2(\omega\times]0,T[)}-\alpha_{i_0}\right\}\tag{A.3.1}$$

where $|\ .\ |_H$ denotes the norm in $L^2(\omega\times]0,T[)$. We now note that the quantity

$$\left|L^*(g,\alpha)\right|^2_{L^2(\omega\times]0,T[)}\tag{A.3.2}$$

defines a norm on $L^2(\Omega\times R^N)$. In fact, if $L^*\{g,a\}=0$ then

$$\varphi=0 \text{ on } \omega\times]0,T[.\tag{A.3.3}$$

But from (2.12)

$$\varphi'+A\varphi=0 \text{ in } \Omega-\left\{a_i\right\}\tag{A.3.4}$$

(Ω from which one takes out the points a_i) and as $\Omega-\left\{a_i\right\}$ is connected (it is here that a technical difficulty occurs if Ω is of dimension 1), it results from the MIZOHATA [1] uniqueness theorem that

$$\varphi=0 \text{ on }.\left[\Omega-\left\{a_i\right\}\right]\times]0,T[.\tag{A.3.5}$$

Therefore the support of φ is in $\{a_i\}\times]0,T[$, and then $\varphi + A\varphi$ necessarily contains distributions at an order larger than order 0 of $d(x-a_i)$. Hence $j=0$ in $\Omega\times]0,T[$ so that $g=0$ and $\alpha_i = 0$.

Therefore if we pose

$$\|\{g, a\}\|_F = \left|L^*(g,\alpha)\right|_{L^2(\omega]0,T[)} \tag{A.3.6}$$

we define a (new) norm on $L^2(\Omega)\times R^N$ and we denote F the Hilbert space completed of $L^2(\Omega)\times R^N$ for that norm. The problem (A.3.1) then is identical to

$$\inf \left\{\frac{1}{2}\left|L^*(g,\alpha)\right|_F^2 - \alpha_{i_0}\right\} \tag{A.3.7}$$

Existence and uniqueness of the solution in F at once result from (A.3.7). It remains to try and make more precise this space F. Since $\alpha \in R^N$, the only difficulty in evidence bears on g. We therefore can restrict ourselves to the case where $\alpha = 0$ and consider the following situation. We define Ψ by

$$\begin{cases} \psi' + A\psi = 0 \text{ in } Q \\ \psi(x,0) = g(x) \text{ in } \varsigma \\ \dfrac{\partial \psi}{\partial v_A} = 0 \text{ in } \Sigma \end{cases} \tag{A.3.8}$$

and we pose

$$\|g\|_G = \|\psi\|_{L^2(\omega]0,T[)}.$$

which endows $L^2(\Omega)$ with a new norm. We denote by G the space completed of $L^2(\Omega)$ for the norm $\|g\|_G$. We then have

$$F = G\times R^N$$

and the solution to (A.3.7) corresponds to $g\in G$.

The space G is a very "large" space. It is the "largest" space for which a generalized solution y of (A.3.8) is such that

$$\chi_\omega\psi \in L^2(\omega\times]0,T[) \tag{A.3.11}$$

This space is not, in general, a space of distributions. It contains, in fact, infinite order distributions outside $\overline{\omega}$, these infinite order singularities being able to be "regularized" by the solution of (A.2.8) (assuming the coefficients of A very regular). The space G is a space of distributions only in the case where $\omega = \Omega$.

Remark A.3.1

Of course, what has just been said is good if g belongs to a finite - dimensional space, in which case the completed space for the norm (A.3.9) evidently is the same finite dimensional space.

References.

I. EKELAND and R. TEMAM [1] Analyse convexe et problèmes variationnels Dunod-Gauthier Villars-Paris, 1974.

J.L. LIONS [1] Sentinels with Special Sensitivity,.IFAC Symposia Series, EL JAI and M. AMOUROUX Eds, Pergamon Press, 1990, p.1-4.

J.L. LIONS [2] Sentinelles pour les systèmes distribués à données incomplètes, R.M.A. Masson_t.21_1992.

S. MIZOHATA [1] Unicité du prolongement des solutions pour quelques opérateurs différentiels paraboliques, Mem.Coll.Sc.Univ.Kyoto. 31(3) 1958 p. 219-239.

T.R. ROCKAFELLAR [1] Duality and stability in extremum problems involving convex functions. Pac. J.of Math., 21, (1967), p.167-187, 1967.

References

B.E. AINSEBA [1], Exact Controllability , Identifiabilité, and sentinels. Ph.D. thesis, Compiègne University of Technologie, 1992.

B.E. AINSEBA, J.P. KERNEVEZ and R. LUCE [1], Identification de paramètres dans des problèmes nonlinéaires à données incomplètes RAIRO, Vol. 28, n°3, 1994, p.313 à 328 .

B.E. AINSEBA, J.P. KERNEVEZ and R. LUCE [2], Application des sentinelles à l"identification des pollutions dans une rivière, RAIRO, Vol. 28, n°3, 1994, p.297 à 312.

B.E. AINSEBA, J.P.KERNEVEZ and R. LUCE, [3],Identification of parameters in nonlinear problems with missing data. Application to the identification of pollutions in a river, p590-599 in"J.Henry and J.P. Yvon (Eds.) System Modelling and Optimization, Proceedings of the 16th IFIP-TC7 Conference", Compiègne, France, 1993.

D. ACHELI, Identification of the path followed by a polluting ship in the sea and the quantity of pollutant deversed, Ph.D. thesis, Compiègne University of Technologie, 1996.

M.P. ANDERSON and W.W.WOESSNER [1], Applied Groundwater Modeling, Simulation of flow and advective transport, Academic Press, 1992.

Y. BARD, *Nonlinear Parameter Estimation*, Academic Press, New-York, (1974).

J.V; BECK, B. BLACKWELL, AND C.R. ST CLAIR Jr, Inverse heat conduction Ill-posed problems, John Wiley & Sons, New-York, (1985).

J.BEAR [1] *Hydraulics of Groundwater*, Mc Graw-Hill, New York, 1979

E.A.BENDER [1] *An introduction to mathematical modelling.* Wiley, 1978

A. BERMUDEZ and M.E. VAQUEZ [1] Flux-Vector and Flux-Difference Splitting Methods for the Shallow Water Equations in a Domain with variable Depth, p.256-267 in*Computer Modelling of seas and Coastal Regions*, Elsevier Applied Science, P.W. Partridge, Edr.

0. BODART [1] Sentinelles, Thèse, Université de Technologie de Compiègne, 1992.

0. BODART and J.P. KERNEVEZ [1], Sentinels in Rivers, p.69-76 in "Jornadas Hispano France sas sobre Control de Sistemas distribudos", Grupo de Analisis Matematico Aplicado de la Universitad de Malaga, Malaga, Octubere 1990.

0. BODART, J.P. KERNEVEZ and T. MANNIKKO [1], Sentinels for Distributed Environmental Systems, in "Proceedings of the 11th IASTED International Conference on Modelling, Identification and Control", Innsbruck, Austria, February 10-12,1992.

H.D. BUI, *Introduction aux problemes inverses en mécanique des matériaux*, Eyrolles, Paris, (1993).

H.D. BUI, editor, [1] Proceedings of the second international Symposium on inverse problems - ISIP 94 Paris France 2-4 November 1994, Inverse problems in Engineering Mechanics.

G. CHAVENT[1] A unified physical presentation of mixed, mixed hybrid finite elements and standard finite difference approximations for the determination of velocities in waterflow problems, Adv. in Water Resources, Vol. 14, N°6, pp. 329-348.

G. CHAVENT[2] generalized Sentinels Defined via Least Squares, Appl Math Optim 31: 189-218 (1995).

G. CHAVENT [2], Generalized Sentinels Defined Via Least Squares. Rapport de Recherche INRIA N°1932 (1993).

G. CHAVENT [3], Estimation de parametres distribués dans des equations aux dérivées partielles, p.361 - 390 in " computing Methods in Applied Sciences and Engineeering", Part 2, Lecture notes in Computer Science 11, Springer-Verlag, New York 1974.

G. CHAVENT [4], Analyse fonctionnelle et Identification de coefficients répartis dans les équations aux dérivées partielles, Thesis, Paris, 1971.

P. DEMESTERRE [1] Sentinelles, Thèse, Université de Technologie de Compiègne, 1996.

/H.W.ENGL and W. RUNDELL, editors, "Inverse problems in diffusion processes", proceedings of the GAMM-SIAM symposium, (1994,: Lake St. Wolfgang), edited by GAMM & SIAM, 1995.

R.S.FALK, Approximation of Inverse problems, p.7-16 in "Inverse problems in partial differential equations", D.Colton,R.Ewing and W.Rundell, Editors, SIAM

J.L.FEIKE and J.H. DANE, Analytical solutions of the One-Dimensional Advection Equation and Two- or Three- Dimensional Dispersion Equation, Water Resources Research, Vol. 26, N°7, pp. 1475-1482.

J. GRASMAN [1], Methods for improving the prediction of dynamical processes with special reference to the atmospheric circulation, Technical note 94-01, Department of Mathematics,Waveningen Agricultural University, The Netherlands.

M. Kalivianakis, S.L.J. Mous and J. Grasman, Reconstruction of the seasonally varying contact rate for measles, Technical note 91-06,Department of Mathematics,Waveningen Agricultural University, The Netherlands.

W.KINZELBACK[1] *Groundwater modelling, An Introduction with Sample Programs in BASIC*, Developments in Water Science 25, Elsevier.

M.M.LAVRENTIEV,V.G.ROMANOF and S.P.SHISHAT-SII [1] *Ill-posed Problems of Mathematical Physics and Analysis*, translations of Mathematical Monographs,Volume 64, AMS, Providence, Rhode Island.(1986).

M.M. LAVRENTIEV, V.G. ROMANOF and VASILIEV [1]: *Multidimensional Inverse Problems for differential Equations* Lecture Notes in Mathematics, N° 167, Springer-Verlag, 1970.

F.-X. LE DIMET [1] Methods of Optimal Control for the Assimilation of Data in Meteorolpgy, Ph.D. thesis, University of Clermont-Ferrand, 1990.

F.-X. LE DIMET and O. TALAGRAND [1], Variational algorithms for anzalysis and assimilation of meteorological observations: theoretical aspects, Tellus,(1986), 38A, 97-110.

J.L. LIONS [1]*Contrôle optimal de systèmes gouvernés par des équations aux dérivées partielles.* Dunod, Paris, 1968.

[2] Sur les sentinelles des systèmes distribués. C.R.A.S.Paris,t 307, 1988. Le cas des conditions initiales incomplètes, p. 819-823 Conditions frontières, termes sources, coefficients incomplètement connus, p. 865-870.

[3] Sentinels and stealthy perturbations, International Symposium on Assimilation of Observations in Meteorology and Oceanography, Clermont-Ferrand, July 9-13, 1990.

[4] Furtivité et sentinelles pour les systèmes distribués à données incomplètes, CRAS, Paris, 1990.

[5] Sentinels for periodic distributed systems. Chinese Annals of Math. 10 B (3), (1990), P. 285-291.

[6] "Exact controllability, stabilization and perturbations for distributed systems" S.I.A.M. Rev., 30, n°1, 1988.

[7] Quelques notions sur l'Analyse et le Contrôle de systèmes à données incomplètes, p.43-54 in " Actas del I Congresso de Matematica Applicada, Grupo de Analisis Matematico Applicado de la Universitad de Malaga", Septiembre 1989.

[8] Sur les sentinelles non linéaires dans les systèmes distribués. (En Russe). Troudy Akad Nauk CXCII, Moscou,, 1990, p. 140-145.

[9] Sentinels with special sensitivity, IFAC Symposia Series, El Jai et Amouroux Edrs, Pergamon Press N°3-1990, p.1-4.

[10] *Sentinelles pour les Systèmes Distribués,* Masson, 1992.

Y. LIU [1] Algorithmes pour la méthode des éléments finis et pour la méthode de continuation: application à la contrôlabilité exacte, Thèse, Compiègne, (1989).

R.LUCE [1] Contrôlabilité exacte de systèmes régis par des équations aux dérivées partielles. Problèmes elliptiques et paraboliques, linéaires ou non linéaires, problème de Rayleigh-Bénard, Thèse, Compiègne (1990).

T.MANNIKO, O.BODART, and J.P.KERNEVEZ, Numerical methods to compute sentinels for parabolic systems with an application to source terms identification, p.670-679 in"J.Henry and J.P. Yvon (Eds.) System Modelling and Optimization, Proceedings of the 16th IFIP-TC7 Conference, Compiègne, France, 1993.

J.F.C.A. MEYER and S.E. PALOMINO CASTRO, Mathematical modelling and numerical simulation of an air pollution problem and its local effects, p.33-40 in "Computer Techniques in Environmental Studies V, Vol. I: Pollution Modeling, Editor: P. Zannetti", Computational Mechanics Publications, Southampton and Boston,1994.

P.MELLI and P.ZANNETTI[1] *Environmental Modelling*, Elsevier, 1992.

S.MIZOHATA [1], Mem.Coll.Sc.Univ.Kyoto, A 31, (3), 1958, p.219-239.

H. NINOMIYA and K. ONISHI[1] *Flow Analysis Using a PC*, Computational Mechanics Publications,Southhampton Boston, 1991.

R. MOSE, P.SIEGEL and Ph. ACKERER, Simulation des écoulements en milieu poreux par des éléments finis mixtes hybrides, Hydrogéologie, N°4, pp. 293-302, 1993.

OKUBO [1] *Diffusion and Ecological Problems:Mathematical Models*, Springer-Verlag, 1980.

J.M.ORTEGA and W.C.RHEINBOLDT *Iterative solution of nonlinear equations in several variables*, Academic Press, New York and London, 1970

M.S.PILANT and W.RANDEL, Undetermined Coefficient Problems for Quasilinear Parabolic equations, p.165-185 in "Inverse problems in partial differential equations", D.Colton,R.Ewing and W.Rundell, editors, SIAM

C..POULARD [1] , Identification de la contribution de plusieurs pollueurs à la pollution globale d'un aquifère par la méthode des sentinelles, Mémoire présenté pour l'obtention du Diplôme d'Ingénieur de l'Ecole Nationale du Génie de l'Eau et de l'Environnement de Strasbourg et du Diplôme d'Etudes Approfondies "Mécanique etg Ingénierie", Filière:" Sciences de l'Eau", Strasbourg, 1994.

C.POULARD R. MOSE and Ph. ACKERER, Application of the sentinel method to determine the contribution of pollution sources in an aquifer. Journées Numériques de Besançon, 1994

K.SPERRY[1] The Battle of Lake Erié: Eutrophication and Political Fragmentation. Science, Vol. 158, pp. 351-355, 1967

R.H.RAINEY,[1] Natural Displacement of Pollution from the Great Lakes. Science, Vol. 155, pp. 1242-1243, 1967

H.W. STREETER and E.B.PHELPS [1] A study of the pollution and natural purification of the Ohio River, Bull. 146 Public Health Service Washington D.C., 1925.

C.TAYLOR andJ.M.DAVIS [1] Tidal propagation and dispersion in Estuaries, p.95-118 in: R.H.Gallagher, J.T. Oden, C. Taylor and O.C. Zinkiewicz Edrs., *"Finite elements in fluids "*, Volume 1, Wiley, 1975.

R. TEWARSON [1] , Comp.J; 1, 411-413 (1968).

A.N. TICHONOV et V.Ya ARSENIN [1] *Solutions of Ill-Posed Problems*, Winston-Wiley,1977.

F.VAN DEN BERGHE [1] Variational assimilation of remote sensing data for the mapping of pollutant sources in lakes, 12th Symposium of the European Association of Remote sensing Laboratories,7-11 september 1992, Eger, Hungary.

S.G.WALLIS ET D.W.KNIGHT [1]. On Quality Assurance for Numerical Tidal Models, p.149-160 in*Computer Modelling of seas and*Coastal Regions, Elsevier Applied Science,P.W.Partridge, Edr.

CL..WROBEL and C.A.BREBBIA [1]*Water Pollution: Modelling, Measuring and Prediction*, Elsevier, 1991.

Index